+BF444 .S36 1987+

AUDREY COHEN COLLEGE
50664000142366
Scholz, Roland W/Cognitive strategies in
BF444 .S36 1987 C.1 STACKS 1987

BF
444
S36
1987

Scholz, Roland W.
    Cognitive strate-
gies in stochastic
thinking

DATE DUE

AUDREY COHEN COLLEGE LIBRARY
345 HUDSON STREET
NEW YORK, NY 10014

COGNITIVE STRATEGIES IN STOCHASTIC THINKING

THEORY AND DECISION LIBRARY

General Editors: W. Leinfellner and G. Eberlein

> Series A: Philosophy and Methodology of the Social Sciences
> Editors: W. Leinfellner (Technical University of Vienna)
> G. Eberlein (Technical University of Munich)
>
> Series B: Mathematical and Statistical Methods
> Editor: H. Skala (University of Paderborn)
>
> Series C: Game Theory Mathematical Programming and Mathematical Economics
> Editor: S. Tijs (University of Nijmegen)
>
> Series D: System Theory, Knowledge Engineering and Problem Solving
> Editor: W. Janko (University of Vienna)

SERIES A: PHILOSOPHY AND METHODOLOGY OF THE SOCIAL SCIENCES

Editors: W. Leinfellner (Technical University of Vienna)
G. Eberlein (Technical University of Munich)

Editorial Board

M. Bunge (Montreal), J. S. Coleman (Chocago), M. Dogan (Paris), J. Elster (Oslo), L. Kern (Munich), I. Levi (New York), R. Mattessich (Vancouver), A. Rapoport (Toronto), A. Sen (Oxford), R. Tuomela (Helsinki), A. Tversky (Stanford).

Scope

This series deals with the foundations, the general methodology and the criteria, goals and purpose of the social sciences. The emphasis in the new Series A will be on well-argued, thoroughly analytical rather than advanced mathematical treatments. In this context, particular attention will be paid to game and decision theory and general philosophical topics from mathematics, psychology and economics, such as game theory, voting and welfare theory, with applications to political science, sociology, law and ethics.

ROLAND W. SCHOLZ

*Institute for Didactics of Mathematics,
University of Bielefeld, F.R.G.*

COGNITIVE STRATEGIES
IN
STOCHASTIC THINKING

D. REIDEL PUBLISHING COMPANY

A MEMBER OF THE KLUWER ACADEMIC PUBLISHERS GROUP

DORDRECHT / BOSTON / LANCASTER / TOKYO

Library of Congress Cataloging in Publication Data

Scholz, Roland W.
  Cognitive Strategies in stochastic thinking.

  (Theory and decision library. Series A, Philosophy and methodology of the social sciences)
  Bibliography: p.
  Includes index.
  1. Human information processing. 2. Decision-making. 3. Judgment. 4. Stochastic processes. I. Title. II. Series.
  BF444.S36   1987   153.4   87–4385
  ISBN 90-277-2454-7

Published by D. Reidel Publishing Company,
P.O. Box 17, 3300 AA Dordrecht, Holland.

Sold and distributed in the U.S.A. and Canada
by Kluwer Academic Publishers,
101 Philip Drive, Assinippi Park, Norwell, MA 02061, U.S.A.

In all other countries, sold and distributed
by Kluwer Academic Publishers Group,
P.O. Box 322, 3300 AH Dordrecht, Holland.

All Rights Reserved
© 1987 by D. Reidel Publishing Company, Dordrecht, Holland
No part of the material protected by this copyright notice may be reproduced or
utilized in any form or by any means, electronic or mechanical
including photocopying, recording or by any information storage and
retrieval system, without written permission from the copyright owner

Printed in The Netherlands

TABLE OF CONTENTS

ACKNOWLEDGEMENTS   IX

INTRODUCTION

1. TOWARD AN UNDERSTANDING OF INDIVIDUAL DECISION MAKING UNDER UNCERTAINTY

1.1.	Stochastic Thinking as an Object of Psychological Research	1
1.2.	The Plan of the Studies	5

PART I

2. THE 'BASE–RATE FALLACY' – HEURISTICS AND/OR THE MODELING OF JUDGMENTAL BIASES BY INFORMATION WEIGHTS

2.1.	Variants of Base–Rate Problems	11
2.2.	Normative Solutions and Degrees of Freedom in Task Analysis	14
2.3.	Psychological Explanations of the 'Base–Rate Fallacy'	21
2.4.	A Model of Information Integration in Base–Rate Problems: The Individualized Normative Solution	26
2.5.	Some Hypotheses on the Degree of Base–Rate Consideration	30
2.6.	Experiment A	35
2.6.1.	Subjects	35
2.6.2.	Procedure and Experimental Tasks	35
2.6.3.	Independent Variables and Experimental Design	36
2.6.4.	Dependent Variables and Measures	37
2.6.5.	Results	40
2.7.	Discussion of Experiment A's Results	52
2.8.	Conclusions	55

3. A CONCEPTUALIZATION OF THE MULTITUDE OF STRATEGIES IN BASE–RATE PROBLEMS

3.1.	Ways of Understanding and Treating Stochastic Problems	57

TABLE OF CONTENTS

3.2.	The Analytic and the Intuitive Mode in Stochastic Thinking	60
3.3.	Experiment B	66
3.3.1.	Subjects	66
3.3.2.	Procedure and Experimental Tasks	67
3.3.3.	Methodological Remarks on the Use of Written Protocols	67
3.3.4.	Measures and Rating Procedure	69
3.3.5.	Results	69
3.4.	Summary and Discussion of Experiment B's Results	84
3.5.	Conclusions	89

4. MODES OF THOUGHT AND PROBLEM FRAMING IN THE STOCHASTIC THINKING OF STUDENTS AND EXPERTS (SOPHISTICATED DECISION MAKERS)

4.1.	The Concerns of this Chapter	93
4.2.	The Impact of Problem Framing, Iteration, Sex, and Career Socialization on the Activated Mode of Thought	95
4.3.	Modes of Thought and the 'Base – Rate Fallacy' – Crippled Intuition versus Fallacious Analysis?	100
4.4.	Hypotheses	101
4.5.	Experiment C	102
4.5.1.	Subjects	102
4.5.2.	Procedure and Experimental Tasks	102
4.5.3.	Independent Variables	105
4.5.4.	Dependent Variables, Measures, and Rating Procedure	106
4.5.5.	Results	108
4.6.	Discussion of Experiment C's Results	123
4.7.	An Investigation into Expert's (Sophisticated Decision Maker's) Behavior in Base – Rate Problems	133
4.8.	Experiment D	135
4.8.1.	Subjects	135
4.8.2.	Procedure	135
4.8.3.	Results	136
4.9.	Discussion of Experiment D's Results	138

TABLE OF CONTENTS

PART II

5. STOCHASTIC THINKING, MODES OF THOUGHT, AND A FRAMEWORK FOR THE PROCESS AND STRUCTURE OF HUMAN INFORMATION PROCESSING

5.1.	A Model or Framework for the Process and Structure of Human Information Processing	144
5.1.1.	An Introduction to the Components of the Framework	145
5.1.2.	The Connections Between the Individual Components	165
5.1.3.	Some Remarks on the Framework's Relationship to Other Approaches to a Modeling of Information Processing	167
5.1.4.	Some Remarks on the Frame Concept and on Cognitive and on Situational Framing	168
5.2.	Modes of Thought as Different Activations of the Cognitive System	170
5.3.	The New Perspective and its Value for an Understanding of the Process of Stochastic Thinking	184
REFERENCES		190
INDEX		215

ACKNOWLEDGEMENTS

Acknowledgements for texts that have taken years to prepare are generally difficult, because the large number of individuals who have contributed to them makes it impossible to suitably thank everyone. An acknowledgement becomes almost impossible, in my opinion, when the work has been accompanied by the controversies that to a large extent arise from the different epistemological, disciplinary, and methodological perspectives that one finds in such a difficult and disputed field as "Cognitive Strategies in Stochastic Thinking".

I am committed to first of all thank those who have accompanied the work through all phases from the beginning to end, and were repeatedly able to overcome the stress that arose from both the content−related and personal demands and conflicts. In first place, I must mention cand. psych. Jonathan Harrow, with whom I was able to discuss continuously the content of the work and the clarity of presentation, and who was responsible for editing my rough English, and equally cand. math. Andreas Bentrup, who helped me with his advice and work during the planning, running, and data analysis of the experiments. On looking back, it has become clear to me that this work would not have been possible without the granting of a postdoctorial scholarship and the financial support of the German Research Foundation (DFG).

Through the DFG, it was possible for me to undertake a brief research residence at Decision Research: A Branch of Perceptronics, Eugene, Oregon. I should like to thank the researchers in this group, in particular Baruch Fischoff, Sarah Lichtenstein, and Maja Bar−Hillel, for the many intense but always constructive disputes that, in retrospect, I would not like to have gone without.

The support, stimulation, and criticism was diverse. I should like to thank Prof. M. Irle, Mannheim; Prof. R. Selten, Bonn; and Prof. M. Waller, Bielefeld/Heidelberg for the informative discussions during the initial and end phases of my work, and also Dr. C.M. Allwood, Göteborg; Dr. M. Borovcnic, Klagenfurt; Dr. R. Bromme and Dr. H. Bussmann, Bielefeld; Prof. D. Dörner, Bamberg; Dr. J. Funke, Bonn; Dr. R. May, Berlin; Prof. W. Tack, Saarbrücken; Prof. R. Schaefer, Mannheim; and also my friend

ACKNOWLEDGEMENTS

Dr. Th. May and my colleagues in Prof. H.G. Steiner's team at the IDM who critically commented on sections of my work or on my running plans, and thus contributed to improvements in the manuscript.

I should also like to mention here the controversial discussions with the Bielefeld psychologists Prof. W. Prinz, Prof. H. Heuer, and Dr. O. Neumann, on the sense and limitations of research into higher cognitive processes that is limited to microprocesses, and also on my own approach. My thanks are, above all, for the exact and helpful comments on Chapters 2 to 4, that certainly improved the text despite fundamentally different views on methodological questions in research. I would further like to thank Prof. U. Schulz, Bielefeld, for his comments on methodological problems.

For outsiders, long phases of planning, data collection, and reflection sometimes unexpectedly turn into intensive phases of text production. My thanks are earned above all by Carmen Buschmeyer, Herta Ritsche, and, once again, Andreas Bentrup who managed to cope with these wave−like processes, without being washed over by the crests, or being disheartened by the current difficulties of producing texts on word processors.

Bielefeld, 1986
Roland W. Scholz

INTRODUCTION

1. TOWARD AN UNDERSTANDING OF INDIVIDUAL DECISION MAKING UNDER UNCERTAINTY

1.1. STOCHASTIC THINKING AS AN OBJECT OF PSYCHOLOGICAL RESEARCH

Decision making under uncertainty is one of the ubiquituous enterprises of life. Many situations are characterized by the decision maker's incomplete knowledge about **alternatives** or **outcomes**. Who actually knows all the possible alternative interventions if some technical inventory won't function, for instance, if a car refuses to start? Sometimes the alternatives under consideration are seemingly clear, but there is incomplete knowledge about the outcomes. Even if there is no chance involved in a decision, and the choice alternatives are well defined, there may be substantial uncertainty. Anyone who has been offered a new job or is considering starting up a new business normally has to make his decision in the presence of ignorance about a lot of facilities and thus the outcomes of his decision. Yet even if both the possible alternatives and the outcomes of all alternatives are known, uncertainty about **the probability of** the **occurence** of the **outcomes** may be present. Any owner of stocks and shares knows that they may either rise, or fall, or will remain unchanged during the course of a day. Although there is a lot of information about what may happen, even the experienced stockbroker usually does not know whether specific stocks will rise till closing time from the early dealings, nor does he know the chances of a rise or fall.

In general, decision making under uncertainty is characterized by incomplete knowledge about a situation, which means that the decision maker does not know the possible **alternatives,** the **alternatives' possible outcomes**, or the **probability** of the **occurrence of the** outcomes (cf. LEE, 1977/1971). The sources of uncertainty itself are not homogeneous, thus for instance, the uncertainty may be brought about internally ("I am not sure whether I left my umbrella at John's house") or externally ("It is not certain whether it will stop raining tomorrow") (cf. KAHNEMAN & TVERSKY, 1982a; BERKELEY &

HUMPHREYS, 1982; SCHOLZ, 1983a). Within the context of this text, decision making under risk, in which the alternatives, the potential outcomes, and their probability of occurence are known — but not the actual outcome (for instance when playing blackjack) — will also be subsumed under the heading decisions under uncertainty.

Originating from the controversy on clinical versus actuarial decisions in clinical inference in medical problem solving and other fields (cf. e.g., MEEHL, 1954; MEEHL & ROSEN, 1955; GOUGH, 1962; HAMMOND, HURSCH, & TODD, 1964; SAWYER, 1966; GOLDBERG, 1970; SHULMAN & ELSTEIN, 1975), experimental studies on probability learning (cf. ESTES, 1964, 1976), and studies subsequent to EDWARDS' attempt at the rather direct application of the Bayesian framework as a model of information processing (cf. EDWARDS, LINDMAN, & SAVAGE, 1963; extended reviews on this research are given by PETERSON & BEACH, 1967; SLOVIC, FISCHHOFF, & LICHTENSTEIN, 1977; RAPOPORT & WALLSTEN, 1972), psychological research into decision making has been significantly elaborated by Daniel KAHNEMANN und Amos TVERSKY's seminal work on judgmental heuristics. Their theoretical studies were based on a series of experiments conducted during the seventies. These experiments focused on demonstrations of human biases, fallacies, errors, and shortcomings in probability judgment, choice, and decision behavior. In some respects, KAHNEMAN and TVERSKY's work may be regarded as the beginning of a cognitively oriented decision science which attempts to grasp the (cognitive) psychological foundations of decision behavior in contrast to the descriptive and more quantitively oriented behavioral decision theory.

A key assumption within KAHNEMAN and TVERSKY's line of research (for an overview see KAHNEMAN, SLOVIC, & TVERSKY's, 1982, representative collection of papers, or NISBETT & ROSS, 1980) is the existence of a valid yardstick, point of reference, or measure of comparison for the evaluation of judgment and decision. Usually this yardstick is thought to be provided by stochastic models or principles such as subjective expected utility or hypothesized stochastic independence of events. The validity of this assumption has been repeatedly criticized, for many experimental tasks do not have an unequivocal solution that is provided by a stochastic model (cf. EINHORN & HOGARTH, 1981; SCHOLZ, 1981; BERKELEY & HUMPHREYS, 1982; LOPES, 1982; JUNGERMANN, 1983; ZIMMER, 1983). On the contrary, the individual or laypersons' fallacious heuristics and cognitive strategies in the process of decision making under uncertainty are usually considered to be free of statistic arguments and knowledge about probability.

Within this book, **stochastic thinking** is introduced as a central concept for a modeling and a conceptualization of the cognitive activity in decision making under uncertainty. As the term 'stochastic' currently is used differently in the English speaking world than in Germany and probably the rest of continental Europe, and, as far as the author is aware, it has not been applied in the literature on psychological decision research, the definition given below is preceded by some remarks on the concept of stochastic and its applications outside of psychology.

The different usage of stochastic can be illustrated by comparing dictionaries which provide sufficient and competent information. Whereas, for instance, the Encyclopaedia Britannica, Micropaedia, 1981, presents a rather restricted definition: "a stochastic is a family of random variables ... that for each finite subset ... has a joint probability distribution ...", the "Wörterbuch Philosophie und Naturwissenschaften" provides an alternative broader description: "Stochastic (Greek): the area of science that includes probability theory, statistics, and their application. The object of research is the random dependency of systems and processes. The inclusion of further properties of systems or events, e.g., dependency of time, number, proximity, and other parameters, have, when combined with random events, surpassed the scope and field of classical probability calculus ..." (KRAATZ, 1978, p. 861, our translation). The present study refers to the latter definition when conceptualizing stochastic thinking.

Stochastic thinking has been discussed by European mathematicians and mathematics educators without direct reference to psychological literature. As we consider their arguments to be important for the psychological research on decision making under uncertainty, we shall introduce some of the relevant essentials. In an almost phenomenological approach, involving self−reflection and a self−description of their activity, they emphasize the specificity of cognitive activity in stochastic, which, for instance, is also due to the specific object−relation of stochastic as part of applied mathematics (DINGES & ROST, 1982). Hence RENYI, 1969, p. 85, speaks about the "peculiar form of probability theory thinking" or VARGA, 1972, p. 346, of "probabilistic impregnation of thinking" (cf. STEINBRING, 1980, pp. 372). One of the most pronounced statements on what mathematicians mean by stochastic thinking has been formulated by DINGES, 1977, p. 3: "We believe that there is such a thing as stochastic thinking, perhaps in a similar sense as there is a geometric imaginative faculty (geometrische Vorstellungskraft). One cannot reduce it to the ways of thinking of pure mathematics (just as one cannot reduce geometrical insight to a calculus)" (our translation). A crucial activity in stochastic thinking is the building of stochastic models (cf. HEITELE, 1976, p. 185), or as FREUDENTHAL, 1961, p. 79, has pronounced, "no scientist is as model–minded as the statistician; in no other branch of science is the word model as often and consciously used as in statistics".

With reference to both psychological decision research, which has scarcely dealt with individual model building activity, and the approach to stochastic thinking which has been formulated by mathematicians but has not been experimentally examined, we propose the following definition of stochastic thinking:

Stochastic thinking denotes a person's cognitive activity when coping with stochastic problems, and/or the process of conceptualization, of understanding, and of information processing in situations or problem coping, when the chance or probability concept is referred to, or stochastic models are applied. While in the former variant of stochastic thinking, a subjective stochastic conceptualization is not necessarily needed, the latter does not have to involve stochastic situations. Terms that have received a similar notion are probabilistic reasoning (cf. EDDY, 1982) or statistical inference (e.g., BOROVCNIK, 1982), both of which narrow the scope. However, our definition of stochastic thinking encompasses both concepts.

A closer analysis reveals that some of the concepts included in this definition require further explanation. For instance, what is meant by "the chance or probability concept are referred to", or by "stochastic situation". Some explanation of the former seems to be useful, as there is no unique probability concept, but rather a concept field that contains many facets or components and preconcepts (cf. SCHOLZ & WALLER, 1983). By preconcepts of the individual's probability field, we mean elements of knowledge that refer to certain facets of the individual's probability field, but that are only loosely linked to the (theoretical) core of these facets. For instance, a statement like, "this approach seems more plausible (reasonable)" may be related to logical probability, whereas fuzzy expressions like "more likely" or "probably" may be related to various meanings or facets of probability. But also perceptual intensities or impressions may be a preconcept of, for example, the concept of geometrical probability. However, the question of when a person actually refers to the probability concept cannot receive a general answer, as this depends on the concrete structure of the individual's probability field. When one considers the model—oriented approach in stochastic and also the 'heterocellular' structure of the probability concept, it scarcely seems possible to determine an objective criterion of 'stochasticity' (randomness) in real life situations or even physical processes. For instance, radioactive decay is considered to be a genuine stochastic process in most physics textbooks. However, a minority of renowned theoretical physicists have put forward so—called hidden variable theories (cf. BAUMANN & SEXL, 1984; BRODY, 1983) which suggest alternative explanations. In the following, we will use the concept 'stochastic situation' if there are salient scientific theories or models that refer to the probability concept in the specific situation.

1.2. THE PLAN OF THE STUDIES

In accordance with our view on stochastic thinking in decision making under uncertainty, three different aspects may be emphasized in psychological research: (1) The **behavioral analysis,** which may focus on analyzing or improving the product of the decision process. (2) The **procedural analysis,** which may, for instance, attempt to identify cognitive strategies or heuristics when tracing the process of thinking; and (3) the **semantic** or **conceptual** side, for instance, when analyzing the individuals's knowledge about probability or his/her use of probability to conceptualize uncertainty.

A difference in the emphases on these three aspects is to be found in the various experiments which are reported in Part I of this text. In some respects, the changing accent might reflect the advances in the attempts to accomplish the following major intentions which were: to gain insight into the process of stochastic thinking, in particular the process of probability judgment, and to provide a conceptualization of certain types or classes of cognitive strategies. For instance, Experiment A mainly focuses on the behavioral analysis, although interpretations of the construction principles of the model which we introduced and tested refer to various concepts of probability. Experiments B to D, however, were designed to trace the judgmental process. The results and the insight provided by the experimental studies enable us to sketch a model of the structure and process of the cognitive activity in stochastic thinking and decision making that allows for an integration of different modelings and conceptualizations.

The **experimental paradigm** which was used in the studies in Part I is the so-called **'base-rate-fallacy'**, or to be more precise, variants of KAHNEMAN and TVERSKY's Cab problem (cf. TVERSKY & KAHNEMAN, 1979). There are many reasons for choosing this paradigm in order to investigate salient cognitive activities of stochastic thinking, and this is discussed in **Chapter 2**. In the introduction, we shall refer to only two of these reasons. First, although our experimental paradigm was a well-defined experimental task, it provides an excellent opportunity to discuss the mutually interrelating problems of task analysis (e.g., the role of normative models), of the individual's process of problem structuring, and of the value and limits of cognitive models in a manner which is often only possible when dealing with complex real life situations. Second, many of the fundamental and controversial discussions in psychological decision research deal with the 'base-rate fallacy'.

We will now briefly describe the contents of the chapters and their objectives.

Part I: Chapter 2 introduces the experimental paradigm and discusses the above-mentioned mutual interrelationship of task analysis, problem structuring, and modeling of decision behavior. Furthermore, the traditional standard methodology applied in research into probability judgments is briefly criticized. These considerations lead to the design of an experiment that had to meet two objectives: First, in order to gain insight into the subject's conception of the problem and information processing, information weights were measured, and a model of an individualized normative solution was developed and tested. Second, some independent variables like age, extremity of base-rates, etc. were manipulated in order to test some controversial hypotheses on the differential consideration or neglect of base-rates and the dependence of the model's validity on age. Some interesting findings may already be noted here. For instance, the response behavior of older, higher educated samples was more biased than that of younger subjects, and second, when compared to the 'normal normative solution', the proposed model of an individualized normative solution only revealed an improved prediction for the oldest and most highly educated sample.

Many probability judgments remain unexplained, both by traditional judgmental heuristics and by the proposed model. The object of Experiment B's analysis was subjects' written justifications for their probability judgments. A study of the written justifications revealed that there is a "Multitude of Strategies in Base−Rate Problems". A theoretical frame for an understanding, conceptualization, and classification of this multitude is given by different modes of thought. **Chapter 3** investigates whether an introduced definition by lists of features and attributes of an intuitive versus analytic mode of thought provides a reliable procedure for categorizing strategies in probability judgments.

Experiment B yielded another interesting result. Surprisingly, most of the so−called diagnostic responses (i.e., the total neglect of base−rate information) which have traditionally been considered to be a consequence of intuitive judgmental heuristics (cf. KAHNEMAN & TVERSKY, 1979a, 1982b), were judged to have been processed in an analytic mode! Experiment C was designed to replicate this finding, and to test various hypotheses on the dependency of the elicited mode of thought on situational and differential variables. This is reported in **Chapter 4**. The experiment, which in addition controlled subjects problem understanding, was run with a sample of students. In Experiment D, a sample of ten university professors in statistics and related fields was tested in order to obtain information on experts' intuitive and analytical processing with base−rate problems. Contrary to the students, the university professors' intuitive judgments produced a great number of diagnosticity responses. The differences between students and experts' behavior are discussed with reference to the different knowledge bases of these samples.

Part II: From a theoretical point of view, a definition of modes of thought in stochastic thinking by lists of features seems to be a rather unsatisfactory approach. Furthermore, both the various experimental findings reported in Part I, and the different advances in modeling activities in stochastic thinking still require a theoretical synthesis. **Chapter 5** provides a framework of the "Process and Structure of Information Processing" which allows for a more precise definition of stochastic thinking and provides an approach for integrating the various rather atomistically formulated judgmental heuristics, the proposed individualized normative solution, and other models for probability judgment and activities involved in stochastic thinking. The presented framework also offers a way of describing how the conceptual side (that means the individual's specific interpretation of uncertainty or probability) might be tied to both the procedural side (that means the heuristics and cognitive operations that are actually applied) and also the individual's orientations or goals in stochastic thinking.

PART I

2. THE 'BASE−RATE FALLACY' − HEURISTICS AND/OR THE MODELING OF JUDGMENTAL BIASES BY INFORMATION WEIGHTS

For more than a decade now, the 'base−rate fallacy' has been treated as a theoretical puzzle in both psychology and branches of other sciences concerned with human decision behavior and probability judgments. Generally, the term 'base−rate fallacy' denotes the phenomenon that when people are exposed to additional or diagnostic information about the probability of an event, they often show insufficient consideration of prior actuarial knowledge, hypotheses, or base−rates. The extent of interest in the 'base−rate fallacy' and of research running under this title can be seen from the list of papers which contain extended reviews or the many chapters in books in which the discussion about base−rates is predominant (e.g., AJZEN, 1977; BAR−HILLEL, 1980, 1983; BORGIDA & BREKKE, 1981; GINOSAR & TROPE, 1980; LYON & SLOVIC, 1976; KASSIN, 1979; SCHOLZ, 1981, 1983b; WALLSTEN, 1983; NISBETT & ROSS, 1980).

The 'base−rate fallacy' has been used as a paradigm in challenging controversies over the nature of human rationality (cf. KAHNEMAN & TVERSKY, 1979b; COHEN, 1979, 1981), modes of thought and the framing of problems (cf. SCHOLZ, 1981; BAR−HILLEL, 1983, p. 43), theories of information representation and processing (cf. KAHNEMAN & TVERSKY, 1982a; BIRNBAUM, 1983), and even the probability concept (COHEN, 1981; KYBURG, 1983).

Within experimental psychological work, task analysis has sometimes been neglected. In particular, normative solutions of tasks used in studies on probabilistic problem solving are seldom unique (cf. SCHOLZ, 1981; EINHORN & HOGARTH, 1981; BERKELEY & HUMPHREYS, 1982; PHILLIPS, 1983). This is partially due to the fact that the probability of an event (i.e., unlike the time extension of an event or the length of an object) is much more impressed on a situation, or is a reconstruction of a problem or situation, rather than a straightforward image of the problem. This is why, before turning to descriptions and psychological explanations of the 'base−rate fallacy', we will first specify which types or sets of problems we are considering, and then discuss which degrees of freedom may be inherent to a task analysis and to the normative solution in base−rate problems.

2.1. VARIANTS OF BASE−RATE PROBLEMS

One of the initial illustrations that reveals the problem of base−rate neglect was given by HUFF, 1959. There is anecdotal evidence of many people's startled responses when they are told that the following story is actually true.
> A Rhode Island newspaper proclaimed that in 1957, 10% of all pedestrians killed had crossed an intersection on a green light, while only 6% had crossed on a red light.

The first experimental data reporting the phenomenon of ignoring base−rates in probability judgment were published by KAHNEMAN and TVERSKY, 1973. The problems they used were variants of the well−known **Tom W. problem**. Subjects only received a short and obviously unreliable personality description of one Tom W. They considered it more probable that this person worked in Computer Sciences than in Social Sciences. They did this in spite of their common knowledge that there were a lot more students and workers in Social Sciences than in Computer Sciences at that time. Obviously, the subjects' judgment was influenced by a mere description of a 'computer−buff' stereotype. According to BAR−HILLEL, 1983, this type of task, in which base−rate neglect can be observed, is called the **social judgment paradigm**. Within this paradigm, all the probabilities are usually not given explicitly. That base−rates are factually neglected in the case of the Tom W. problem can be shown by varying the population from which Tom W. is drawn, and thus proving that there are inconsistencies in probability judgments. Within KAHNEMAN and TVERSKY's, 1973, p. 241, procedure, subjects were told that a personality description had been drawn at random from a file with 70 engineers and 30 lawyers. A typical text which was offered read:
> Tom W. is a 45−year−old man. He is married and has four children. He is generally conservative, careful, and ambitious. He shows no interest in political and social issues and spends most of his free time on his many hobbies which include home carpentry, sailing, and mathematical puzzles.

The experimental task was to estimate the probability of Tom W. being an engineer. Subjects' responses were hardly affected by variations of base−rates. Thus, an experimental group with a 70 lawyer and 30 engineer instruction produced a mean probability judgment that was very similar to that of a sample with a 70 engineer and 30 lawyer instruction.

In contrast to problems of the social judgment type, base—rate problems of the so—called textbook paradigm type usually have a (accepted unique) normative solution. A prototype of the **textbook paradigm** is the following **Cab problem** (cf. TVERSKY & KAHNEMAN, 1979):

Cab problem: A cab was involved in a hit—and—run accident at night. Two cab companies, the Green and the Blue, operate in the city. You are given the following data:
(i)   85% of the cabs in the city are Green and 15% are Blue.
(ii) A witness identified the cab as a Blue cab. The court tested his ability to identify cabs under the appropriate visibility conditions. When presented with a sample of cabs (half of which were Blue and half of which were Green) the witness made correct identifications in 80% of the cases and erred in 20% of the cases.
Question: What is the probability that the cab involved in the accident was Blue rather than Green?

The present volume focuses on the **textbook paradigm** of the 'base—rate fallacy'. We will present three more examples in order to enable the reader to gain insight into the structural equivalence of, the differences between, and the ambiguities in variants of the textbook paradigm, and provide some concrete texts which will then be dealt with in the task analysis. We should mention that the identical texts were used in the subsequent experimental studies.

Hit Parade problem: In a South German radio station's hit parade, a studio guest chosen from the listeners' fan mail is asked to predict whether a newly presented song will become a hit, that is, whether a newly presented title will be among the 10 songs most named by the listeners.

One of this program's fans has collected the following data: During the last five years, 35% of the newly presented songs became accepted into the hit parade; 65% of the songs did not.

He has also found out that the studio guests made 80% correct predictions, both for the songs that became a hit and for those that did not.

Now the question: What is the probability that a randomly chosen song that is named by the studio guest as a prospective hit will also be named as a hit by the listeners?

The Motor problem was designed by BAR—HILLEL, 1980, and the TV problem was designed analogously to LYON and SLOVIC's, 1976, light bulb problem.

Motor problem: A large water—pumping facility is operated simultaneously by two giant motors. The motors are virtually identical (in terms of model, age, etc.), except that a long history of breakdowns in the facility has shown that one motor, call it A, was responsible for 95% of the breakdowns, whereas the other, B, caused 5% of the breakdowns only.

To mend the motor, it must be idled and taken apart, an expensive and drawn out affair. Therefore, several tests are usually done to get some prior notion of which motor to tackle. One of these tests employs a mechanical device which operates, roughly, by pointing at the motor whose magnetic field is weaker. In 4 cases out of 5, a faulty motor creates a weaker field, but in 1 case out of 5, this effect may be caused accidentally.

Suppose a breakdown has just occured. The device is pointed at Motor B. The question: What is the probability of Motor B being defective?

TV problem: In a television set factory, screens (cathode ray tubes) are tested by means of an electronic device. If a screen is technically okay, a green lamp lights up, and if it is defective, a red one.

Tests over several years have shown that 75% of the tubes are okay and 25% are defective.

Now it has emerged that the testing device does not work properly, but that there is an additional electrical impulse which can make the red lamp light up in the case of technically faultless tubes and the green one in case of the defective ones. This falsifying impulse occurs in 10% of all tubes tested.

Now the question: If we randomly choose one of the tubes which make the red lamp light up, what is the probability that it is a defective one?

Another type of experimental base−rate problem was introduced in studies by MANIS, DOVALINA, AVIS, and CARDOZE, 1980, and in a similar manner by CHRISTENSEN−SZALANSKI and BEACH, 1982. In these experiments, the base−rate information was experienced sequentially (for instance, by watching slides showing different persons), and the feedback about a questioned event (e.g., a certain disease) was usually also introduced sequentially. We will not go into the detail of these variants of the base−rate problem (critical comments on these experiments are provided by BAR− HILLEL & FISCHHOFF, 1981, and BEYTH−MAROM & ARKES, 1983), but will proceed to describe the structural equivalence of base−rate problems.

In each of the base−rate problems two kinds of information are given. First, there is the base−rate information, which may also be called a priori information or statistical information, as it is often − but not always − known in advance or has been provided by statistical data. In the case of the Tom W. problem, this information may consist of the individual's knowledge about the frequencies of Social and Computer Scientists, or the rates of engineers and lawyers. In the textbook problems presented here, the base−rate information is the rate of Blue cabs (Cab problem), the percentages of titles that actually become a hit (Hit Parade problem), the relative defect rate of Motor A compared to B (Motor problem), and the rate of defective screens (TV problem). As the base−rate information usually provides the starting point in the individual's information processing (which may also be subjectively generated, as for instance in clinical inference, cf. LUSTED, 1968, 1976; MAI & HACHMANN, 1977), we will speak about a **hypothesis, H**.

The second type of information is the indicant or additional information. This is the personality sketch in the Tom W. problem, the witnesses or the studio guests' hit rate in the Cab and Hit Parade problem's or the accuracy of the test device in the Motor and the TV problem. This information or datum will be abbreviated as D. Often, this information provides indications for the probability of a **diagnosis**, both for the hypothesis H and its complementary event $\bar{H}$. Therefore D also stands for **diagnostic information**. In general terms, a diagnosis is a probability distribution (or a statistic of it) over hypotheses. One should note, however, that the above Cab problem and its variants are representatives of a special set of base−rate problems in which both the diagnostic accurracy for H and $\bar{H}$ are equal. Although this special case reveals a critical feature of base−rate problems, till now, it has been the predominant object of experimental psychological research. This is also why we will deal with this problem in detail.

2.2. NORMATIVE SOLUTIONS AND DEGREES OF FREEDOM IN TASK ANALYSIS

We want to note that the considerations below will focus on the textbook paradigm of the base−rate fallacy although they may easily be generalized to the Tom W. problem or other base−rate problems (cf. SCHOLZ, 1981, p. 16). The formal solutions to the above problems can be derived from **BAYES' theorem**. For those who are familiar with the **odds form** of this theorem the solution may be straightforwardly derived from its simplest form; that is:

$$\frac{p(H|D)}{p(\bar{H}|D)} = \frac{p(D|H)}{p(D|\bar{H})} \cdot \frac{p(H)}{p(\bar{H})}$$

In this formula, p(H) denotes the **a priori probability** for the hypothesis, H (e.g., that a Blue cab will be involved in the accident, or that a random title from the hit parade becomes a hit), and $p(\overline{H}) = 1 - p(H)$ denotes the probability of the complementary event, $\overline{H}$. The ratio $p(H)/p(\overline{H})$ is named the **a priori odds**.

The likelihood ratio

$$\frac{p(D|H)}{p(D|\overline{H})}$$

provides the information on a piece of evidence (datum) or diagnosis D under H and $\overline{H}$ (e.g., the accuracy of the witnesses testimony for Blues and Greens, or the matching rates of the studio guest for hits and non−hits).

Because $p(H|D) = 1 - p(\overline{H}|D)$, the requested posterior probability $p(H|D)$ (e.g., the probability for a Blue cab given the witnesses testimony that the cab is "Blue", or the probability that a randomly chosen song that is named by the studio guest as a hit will actually become a hit) can be calculated.

For those who do not have BAYES' formula at their disposal, $p(H|D)$ may be derived in the following way. According to the formula of conditional probability,

$$p(H|D) = \frac{p(H \cap D)}{p(D)}$$

As $p(H \cap D) = p(D|H) \cdot p(H)$
and $p(D) = p(D \cap H) + p(D \cap \overline{H})$, the subsequent equation holds:

$$p(H|D) = \frac{p(D|H) \cdot p(H)}{p(D \cap H) + p(D \cap \overline{H})}$$

If the formula of conditional probability is applied for

$$p(D \cap H) = p(D|H) \cdot p(H) \text{ and}$$
$$p(D \cap \overline{H}) = p(D|\overline{H}) \cdot p(\overline{H}),$$

then the normal version of BAYES' formula results:

$$p(H|D) = \frac{p(D|H) \cdot p(H)}{p(D|H) \cdot p(H) + p(D|\overline{H}) \cdot p(\overline{H})}$$

The base−rates $p(H)$, $p(\overline{H}) = 1 - p(H)$, and the diagnosticity $p(D|H)$ and

$p(D|\bar{H})$ are explicitly given in the textbook paradigm of the base—rate problems. Hence $p(H|D)$ may be calculated. In case of the original Cab problem let Blue denote that the color of the cab was actually blue and "Blue" the testimony that the cab was blue, then this yields

$$p(H|D) = p(Blue|"Blue") = \frac{.8 \cdot .15}{.8 \cdot .15 + .2 \cdot .85} = .41$$

For the above—mentioned Hit Parade problem, the normative solution is .68, for the Motor problem .17 and for the TV problem .75. One may generalize the BAYES' formula to more events, for instance, more colors in the Cab problem. If these events are denoted as $H_1, H_2, ..., H_n$, one obtains analogously for any specific event H_i:

$$p(H_i|D) = \frac{p(D|H_i) \cdot p(H_i)}{\sum_j p(D|H_j) \cdot p(H_j)}$$

If we accept OLSON's, 1976, distinction between superficial task characteristics and underlying problem characteristics, we have to question whether the proposed Bayesian solution is an (the) adequate model of the problem characteristic of the above variants. Mapping between words and mathematics is very delicate, as SHUBIK (cf. SCHOLZ, 1981) has pointed out. Hence, even in the case of obviously clear—cut problems, deeper analysis might raise more ambiguities than one would expect. The ambiguity within text interpretation with respect to formal systems is not only inherent to probabilistic inference tasks, but can be observed in fundamental logical relations like the if—then relation. As BRAINE, 1978, and others have pointed out, subjects often do not interpret the if—then relation as truth functional conditional. WASON and JOHNSON—LAIRD, 1972, p. 92 (cf. also DAWES, 1975, p. 126) stated that "if" sometimes "is not a creature of a constant hue", but that its meaning varies depending on its context. Let us briefly check whether this **chameleon theory** may also hold for the Cab problem and its variants.

A critical aspect of written problem tasks is the **incompleteness** or indeterminacy of the description of a real setting. The cues and parameters of real life problems are unlimited. It is rare to have a unique solution in real life problems, and in order to attain such a unique solution, idealization and restriction to a few important features is necessary.

When confronted with the Cab problem, a subject may, for instance, introduce more or less sophisticated assumptions about the business or activity rates of the Blue and the Green companies that are compatible with the given text. These assumptions may lead to modifications of base−rates, and hence this may produce, even within the frame of Bayesian calculus, a reasonable probability $p'(D|H) \neq p(D|H)$.

Adding compatible new information may be perceived as a special case of a phenomenon called **introducing a theory about the information**. We shall discuss three examples of this in some detail in order to demonstrate how the story parameters may be altered in a reasonable interpretation of the given base−rate problem. First, we will show how assumptions may be introduced about the dependency of the diagnosticity, second, an example will be presented to indicate how the diagnosticity may be integrated into the base−rates, and third, general considerations including two concrete illustrations of the reassessment of base−rates via logical or subjective probabilites will be provided.

Example 1: The potential dependency of diagnosticity on base−rates. BIRNBAUM, 1983, discovered that the normative solution to the original version of the Cab problem sketched above conceals a very unrealistic assumption about the nature of the witnesses perceptual abilities. According to the classical (and widely accepted) framework of signal detection theory (cf. GREEN & SWETS, 1966), one may assume that: (1) each color produces a probability distribution of responses within the blue−green continuum, and (2) that a human information processor decides according to a criterion e (within the blue−green continuum) above which, for example, a color is judged to be green. So much for the frame of this theory. The unrealistic assumption is that the ratio of hit rate to false alarm rate is independent of the proportion of cab colors. According to BIRNBAUM, many signal detection experiments show evidence to the opposite, and it was "shown that" even "the answer 80% is compatible with Equation 1" (i.e., BAYES' formula)"and signal detection theory" (cf. BIRNBAUM, p. 91).

As BIRNBAUM notes, the independence of the diagnosticity from the base−rates is nonrealistic in many other stories, such as the so−called Light Bulb problem. In this problem, which is very similar to our TV problem, the subject knows the probability of a defective bulb, and the hit and false alarm rates of a light bulb tester. "Suppose", argues BIRNBAUM, "the light bulb tester measures the current in each bulb for a given voltage. Suppose among bulbs that light, the distribution of current readings is normal for both good bulbs and defective bulbs, but the means of the distributions differ. It seems reasonable that any profit−oriented factory would adjust the criterion for deciding 'defective' as a function of the costs of false alarms (discarding a good bulb), of misses (replacing a bad one), and the probability of a defective bulb."

Example 2: Integrating diagnosticity into base−rates. There is another degree of freedom in the task interpretation of the TV problem. The statement, "Tests over several years have shown ..." in the TV story may be interpreted in two different manners. First, one may suppose that 75% of the screens were okay. Second, one could reasonably assume on the basis of the TV story that the base−rates are determined by the "tests over several years" with the defective scanner. If "G" denotes that the test device shows a green and "R" a red light, and if G denotes that a screen is okay and R defective, then the following equations hold for a screen drawn at random, as 75% are named "G":

$$p("G") = p("G" \cap G) + p("G" \cap R) = .75$$
$$p("R") = p("R" \cap R) + p("R" \cap G) = .25$$

According to the diagnostic information, 90% of all nondefective and 10% of all defective screens are identified as "G". Hence:

$$p("G") = .9 \cdot p(G) + .1 \cdot p(R) = .75$$
$$p("R") = .9 \cdot p(R) + .1 \cdot p(G) = .25$$

Solving for the unknown base−rates, we obtain $p(G) = .81$ and $p(R) = .19$. Starting from these parameters, BAYES' theorem provides,

$$p(H|D) = p(R|"R") = \frac{p("R" \cap R)}{p("R" \cap R) \cdot p("R" \cap G)}$$
$$= \frac{.9 \cdot .19}{.9 \cdot .19 + .1 \cdot .81} = .679$$

which is different from the .75 value resulting from a base−rate of .25 for defective screens.

Example 3: Reassessing base−rates and diagnosticity via logical or

subjective probabilities. Another possibility of transforming the parameters of base−rate problems may be given by reassessing the values according to different theories of probability. The probability concept is by no means homogeneous, and different concepts can be distinguished (GOOD, 1959; HACKING, 1975; SCHOLZ & WALLER, 1983; COHEN, 1979). Whether a meaning, or which meaning is attributed to the probability in the above base−rate problems may be essential for the answer which is given. At this point, we will briefly refer to the concept of logical probability and to subjective probability.

John Maynard KEYNES, 1921/1973, p. 3, defined probability as "various degrees of rational belief about a proposition which different amounts of knowledge authorize us to entertain".

Similarly, within the frame of subjective probability theory, "subjective probability is one's degree of belief based on an evaluation making the best use of all the information available to him and his own skill" (DE FINETTI, 1974). Both definitions explicitly make allowance for transformations of the information given in word problems. The evaluation process may, for instance, consist of combining more or less sophisticated preassumptions with the given information. However, uncertainty or fuzziness about the validity of the base−rate or diagnosticity parameters may already affect the subjective (normative) solution. We will give two simple illustrations.

Illustration 1: When dealing with the Cab problem, a subject may assume that the witnesses credibility does not always exactly equal .8. For the sake of simplicity, we fictionally suppose that our subject has a discrete (subjective or personal) distribution, and he or she assumes the witnesses credibility to be .5 with probability .1, .8 with .6, and even .9 with .3. Then the expected diagnosticity value is still .8 as stated in the original Cab problem, but the expected value for the requested posterior probability $p(H|D)$ is $.1 \cdot .15 + .6 \cdot .41 + .3 \cdot .61 = .45$ which is a little bit higher than .41. This illustration is a special case of the theorem that when keeping the expected value constant, a greater variance in the diagnosticity distributions yields a higher a posteriori probability in the given Cab problem. Perhaps the probability distributions in the illustrations make a rather artificial impression, but it should be noted that symmetric fuzziness, that means any symmetric probability distributions of the witnesses credibility, already leads to a different solution than the 'normal normative solution'.

Illustration 2: We fictionally suppose the witnesses credibility to be .5 with probability .25, and .8 with probability .75. The base-rates are taken as truth. Then the expected value is .75 · .41 + .25 · .15 = .345 which is considerably lower than .41 and reflects the lower mean of the individual diagnosticity distribution.

Example 4: Indeterminacy of base-rates due to theories about their genesis. Another striking example of the relative (in)determinacy of base-rates is attributed to KAHNEMAN and TVERSKY, and has been documented by BAR-HILLEL, 1983. A doctor working in an emergency ward of a large urban hospital knows (from experience) that the base-rate of stomach disorders to heart trouble is higher during holiday weekends than during ordinary weekends. Now, one day a patient comes in complaining of various troubles. The symptoms look to the doctor like those signalling the onset of heart trouble, but the doctor is aware that a mere stomach disorder may sometimes exhibit the same pattern.

Now a first theory could assume that the increase of patients with stomach disorders is due to the nonavailability of family doctors etc. who are on holiday. However, it might be reasoned that, anyhow, everyone with symptoms like those of the patient will be sent to the hospital, and that the base-rate is therefore irrelevant.

In a second theory, one may hypothesize that the danger of stomach disorders is due to different eating and drinking habits, and hence one should be aware of an increase in the number of false alarms, i.e., one should take more account of the probability of patients' symptoms being produced by stomach disorders.

BAR-HILLEL pointed out that both theories may be justifiable. Clearly according to the first theory, the doctor should ignore the base-rates whereas "if ... one believes the latter account, then the population of weekend diseases is truly distributed differently than the weekday population, and doctors would be wise to take that into consideration (as they would take, say, the patient's sex into consideration, if the two sexes have different rates of heart trouble)" (BAR-HILLEL, 1983, p. 58).

The examples may be regarded as demonstrations of the significance of the degrees of freedom which are involved in the modeling process of stochastic thinking. The Bayesian solution, which was introduced earlier, may be regarded as a kind of 'normal' or 'standard normative' solution. Presumably this solution will be most commonly accepted and might be considered as an adequate solution. However, the examples have shown that other solutions may also be considered as the product of a reasonable text understanding and solution process. Hence, we use semiquotes when the standard 'normative solution' of a base—rate problem is mentioned. Clearly, not all responses deviating from the 'normative solution' may thus be regarded as fallacious if the following definition of a fallacy is referred to.

> "A **fallacy** is the result of a cognitive process which, on the basis of information represented in memory, has led to a wrong conclusion, or to a wrong decision. A fallacy may consist of both the application of an inadequate model which has led to a deviating solution instead of a (possible) existing definite formal solution, or of the application of rough rules of inference (such as intuitive estimates) used as a substitute, which systematically result in an inadequate, or incorrect solution" (SCHOLZ, 1983a, p. 5).

As there often is no definite normative solution to base—rate problems, and as there are several ways of taking base—rates into account, not only the term 'normative solution', but also the term 'base—rate fallacy' should be placed in semiquotes.

2.3. PSYCHOLOGICAL EXPLANATIONS OF THE 'BASE—RATE FALLACY'

As we have already mentioned above, the 'base—rate fallacy' denotes subjects' neglect of base—rate information when they are exposed to diagnostic information. Originally the 'base—rate fallacy' was believed to be a pervasive shortcoming of human performance. Studies on the above Cab problem and its variants have also repeatedly emphasized the "robustness of the base—rate fallacy" (BAR—HILLEL, 1980, p. 215). For instance, TVERSKY and KAHNEMAN, 1979, p. 62, state: "Several hundred subjects have been given slightly different versions of this question (i.e., the question on the probability that a Blue cab was involved — the author). For all versions, the modal and median response was 80%. Thus, the intuitive judgment of probability coincides with the credibility of the witness and ignores the relevant base-rates ...".

More recently, several different theoretical approaches (i.e., heuristics or inferential rules) have been hypothesized to explain when base—rates are taken into consideration, and when not. First of all, the following three approaches will be mentioned: the representativeness heuristic, the causal schema, and the availability heuristic. Although the definitions of these three heuristics are both simple and fuzzy, they are sometimes considered to be the trias of human cognition (cf. CRISTENSEN-SZALANSKI & BEACH, 1984).

- **The representativeness heuristic** states that, "a person who follows this heuristic evaluates the probability of an uncertain event, or a sample, by the degree to which it is (i) similar in essential properties to its parent population; and (ii) reflects the salient features of the process by which it is generated" (KAHNEMAN & TVERSKY, 1972, p. 431, 1982b).
- The **causal schema** was introduced as an explanation for human performance on the Cab type problem, after it had been noticed that base-rates are obviously considered under certain circumstances. According to the causality construct, the following holds: If the base−rate information (i.e., the variables specifying the base−rate) are causally linked with the event for which a probability judgment is required, the base−rates are taken into account (cf. TVERSKY & KAHNEMAN, 1979). In particular, in the Cab problem, the base−rates are causally interpreted (cf. TVERSKY & KAHNEMAN, 1979, pp. 62) if two companies, Blue and Green of equal size, are considered to produce different accident rates.
- An alternative explanation of response behavior is the so-called **availability heuristic**. This heuristic was also introduced into theories of probabilistic information processing by KAHNEMAN and TVERSKY. In general, the availability heuristic suggests that probability judgments are influenced "by the ease with which relevant instances come to mind" (TVERSKY & KAHNEMAN, 1973, p. 207). Base−rates tend to be ignored if they are "remote, pallid, and abstract" (cf. NISBETT, BORGIDA, CRANDALL, & REED, 1976), or if the numbers used in the problems are neither easy to recall nor simple to use in calculation (CARROL & SIEGLER, 1977). On the other hand, base−rates are taken into account if they are vivid, salient, and concrete (cf. NISBETT & ROSS, 1980; WALLSTEN, 1983).
- A fourth construct that is worthy of mention is the **specificity** construct. According to BAR−HILLEL, 1980, causality is but one dimension of internal relevance in base−rate problems. Another dimension that she believes to enhance base−rate usage is specificity. Information is more specific and dominates other information if it refers to a smaller subset. In the case of the Cab problem, the diagnostic information refers to just the one taxi and hence is more specific than the base−rate information, whereas in the Motor problem both the base−rate and the diagnostic

information only refer to the two motors and are thus equally specific.

All the above heuristics primarily focus on an explanation of what the median or modal subject is doing, and should be considered as important contributions to the understanding of the process of cognitive inference in decision making. However, there are also various problems tied to the above conceptualizations. We will not discuss the problems of the constructs in detail (this has been done repeatedly, cf. SCHOLZ, 1981; BAR—HILLEL, 1980, 1983; WALLSTEN, 1983). Nevertheless, we will specify some deficiencies in the previous heuristic research into probability judgment.

From a theoretical perspective, the most severe criticism is that the different heuristics are defined in isolation from each other, side by side, without any theoretical considerations about their mutual interdependence. Both a general integrating model of cognitive activity and selection rules determining which heuristic is consulted are still lacking.

Furthermore, as GRONER, GRONER and BISCHOF, 1983, p.101, have pointed out, the **heuristics** in the field of decision and judgment itself are **fuzzily defined** and show but few characteristics of the kind used in cognitive psychology (cf. DUNCKER, 1935), problem solving (cf. POLYA, 1949), or artificial intelligence.

The issues that have been particularily named by GRONER et al. and which are often missing in research on probability judgments are: (1) the representation of problems and their solutions in terms of information processing, (2) the restriction of the search space to a manageable size by means of heuristics, (3) the sufficiency, and (4) the efficiency criteria.

In order to illustrate and specify some of the criticisms, we introduce Figure 1 which presents a schema of the typical strategies in research on probabilistic problem solving.

We have already discussed some of the **problems confounded** with **letter codified problem tasks** and modeling in some detail in Chapter 2.2. (i.e., the indeterminacy of the solution, for example, by introducing a theory about the information). These problems are located on the right hand side of Figure 2.1.

Another central deficiency, however, is to be found in the left path of Figure 2.1. Typically, "research has been focused on data which reflect only the end product of the decision processes ..." (PAYNE, BRAUNSTEIN, & CARROLL, 1978, p. 17). For an analysis of human inference and cognition, both the subject's understanding or **internal representation** of the text, his/her goals (cf. GRONER, et al. 1983), and the **individual information processing** and conclusions, should be indispensible objects of study. Whether information is causally interpreted, or whether probabilities are transformed or weighted, are, for example, empirical questions. However, in the literature that has been cited, neither text understanding, subjective processing, nor the weighting of information are controlled. Even researchers who have themselves conducted series of experiments on heuristics in probability judgments now admit: "In the study of judgment under uncertainty, there have been relatively few attempts to scale directly the properties of the stimuli. Rather, investigators have tended to rely on the face validity of manipulations for such characterizations" (FISCHHOFF & BAR–HILLEL, 1984, p. 403).

A striking example that reveals the shortcomings of both the potential circuits of the strategy of arriving at conclusions about the process solely by observing the product of inference, and shows the need for an increased rigor in definition is given by LA BREQUE, 1980. In a popular introduction to the representativeness heuristic, he presents the following problem:

"Two American psychology professors are friends with a third professor who likes to write poetry, is shy, and is small in stature. What is his field, Chinese studies or psychology?"

LA BREQUE then continues: "According to Nisbett and Ross, people who guessed Chinese studies 'probably were seduced by the representativeness heuristic'. They decided that the personality profile matched their idea of a Sinologist, not that of a psychologist. But they ignored what is more relevant: the odds are that the third professor is a psychologist, because there are far more psychologists than Sinologists in the population, and because the two psychology professors are much more likely to know other psychologists than they are to number a Sinologist among their friends." The careful reader will already have noticed that the conclusion "psychologist" might be drawn **according** to part (ii) of the definition of the representativeness heuristic (cf. p. 29). If two professors of one faculty know a third, the latter very often will be a member of the same staff; thus, of course, this professor is very likely to be a psychologist as well. Without inquiring into the process of inference in the given example, the representativeness heuristic can be applied arbitrarily post hoc in order to 'explain' both a Chinese studies and a psychology response. The distinction between judgment **of** representativeness and judgment **by** representativeness recently discussed by BAR–HILLEL, 1984 (cf. KAHNEMAN & TVERSKY, 1982b), also does not answer the question as to which inference is actually drawn. Besides, in the LA BREQUE story both

CHAPTER 2.3.

Researcher's side

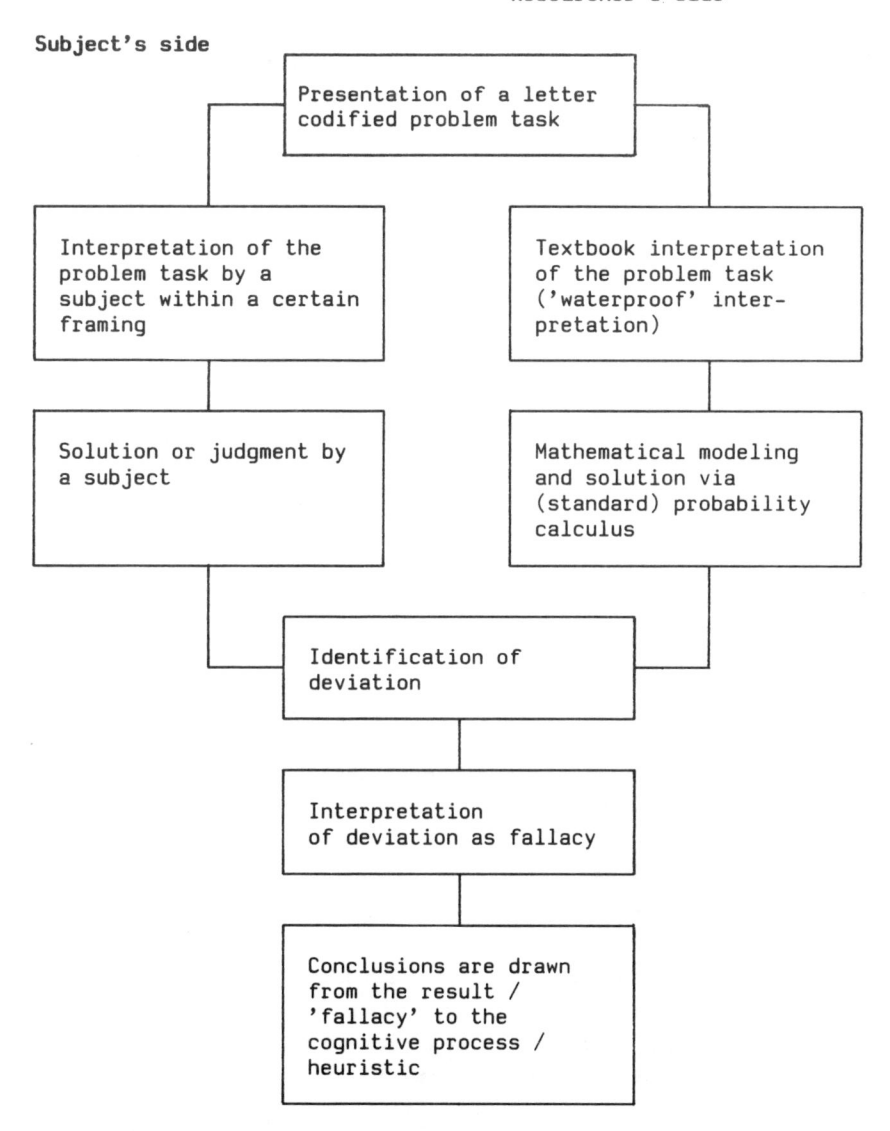

Subject's side

```
                    ┌─────────────────────────┐
                    │ Presentation of a letter│
                    │ codified problem task   │
                    └─────────────────────────┘
        ┌──────────────────────┴──────────────────────┐
┌───────────────────────┐                  ┌──────────────────────────┐
│ Interpretation of the │                  │ Textbook interpretation  │
│ problem task by a     │                  │ of the problem task      │
│ subject within a      │                  │ ('waterproof') inter-    │
│ certain framing       │                  │ pretation)               │
└───────────────────────┘                  └──────────────────────────┘
           │                                          │
┌───────────────────────┐                  ┌──────────────────────────┐
│ Solution or judgment  │                  │ Mathematical modeling    │
│ by a subject          │                  │ and solution via         │
│                       │                  │ (standard) probability   │
│                       │                  │ calculus                 │
└───────────────────────┘                  └──────────────────────────┘
              └──────────────────┬──────────────────┘
                      ┌────────────────────────┐
                      │ Identification of      │
                      │ deviation              │
                      └────────────────────────┘
                                  │
                      ┌────────────────────────┐
                      │ Interpretation         │
                      │ of deviation as fallacy│
                      └────────────────────────┘
                                  │
                      ┌────────────────────────┐
                      │ Conclusions are drawn  │
                      │ from the result /      │
                      │ 'fallacy' to the       │
                      │ cognitive process /    │
                      │ heuristic              │
                      └────────────────────────┘
```

Figure 2.1.: Schema of typical research strategy within probabilistic problem solving and decision making

explanations seem to be judgments of representativeness.

A further shortcoming consists in the lack of control of subjects' actual understanding or decoding of the probability question in the textbook paradigm. For instance, a possible (mis)interpretation of the text may consist of confounding, in the Cab problem and its variants, the question on the probability $p(H|D) = p(Blue|"Blue")$ with the inverse, $p(D|H) = p("Blue"|Blue)$, perhaps since both may be considered to be "natural formalizations" (cf. BAR−HILLEL, 1980, p. 220, 1983, p. 43) of the question in the Cab problem. Although the need for an investigation of subjects' understanding has been recognized, empirical research into this problem is still missing, as far as we know. We will view this problem in detail in Chapters 3 and 4.

2.4. A MODEL OF INFORMATION INTEGRATION IN BASE−RATE PROBLEMS: THE INDIVIDUALIZED NORMATIVE SOLUTION

In the recent study of the 'base−rate fallacy', most of the experiments have been concerned with the investigation of factors which do or do not affect subjects' response behavior within the base−rate problems (for instance varying the causality, etc.). The present situation is that we have some knowledge about how the modal and perhaps the median subject is responding, but we have very little idea about what the rest of the subjects are doing, and how they assess their probability judgments. This remainder is by no means a negligible number. If one looks at the few studies which provide information about the whole response distribution (cf. BAR−HILLEL, 1980; BIRNBAUM & MELLERS, 1983), one can infer that this remainder generally comprises more than half of the subjects.

In order to substantiate our statement about the variety of responses, we shall display two response distributions (cf. Figure 2.2) which have been reconstructed from BAR−HILLEL, 1980, and BIRNBAUM and MELLERS, 1983. The response distribution of BAR−HILLEL's study was obtained in the above Motor problem using a small Israeli sample and has been quoted as a proof for the specificity construct. In the data only 13% so−called diagnosticity responses are to be found. In base−rate problems, a response is named a diagnosticity (base−rate) response if it is identical with the diagnosticity (base−rate). The base−rate parameter is .15 in the BAR−HILLEL, and .30 in the BIRNBAUM and MELLERS study. In both studies the diagnosticity

CHAPTER 2.4.

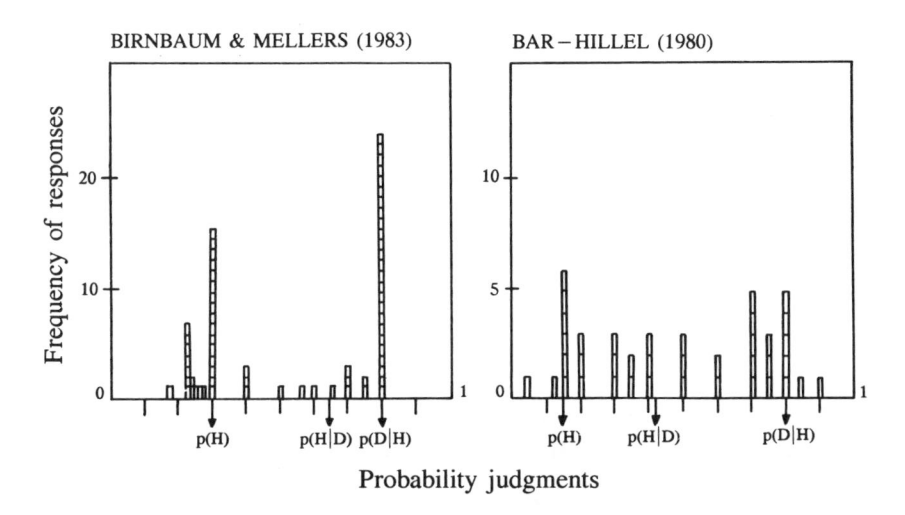

Figure 2.2.: Response distributions from two experiments on the 'base–rate' fallacy (see text)

information is .8. The cover of the BIRNBAUM and MELLERS base–rate problem deals with the probability that a car will last for a certain period. The base–rate information given is about the state of the cars at disposal (lasts longer than three years or not), the diagnostic information is on the validity of the source statements on both states of the car ("good shape" vs. "bad shape"), and the requested probability is the probability that the car will continue to function properly for at least three years, given a judgment "good shape" by the source. The undergraduate student sample in this study produced about 36% diagnosticity responses.

We think that two things may be happening to 50 to 80% of the remainder who do not respond with the diagnosticity. Either they might weight the information and reassess the story parameters when keeping to the problem structure, or however, they might do something completely different, for instance, they might actively interpret the information and relate it to already stored knowledge (cf. HUBER, 1983), and hence start from a different understanding of the story than that intended by the experimenter. In order to gain insight into subjects' conception of base–rate problems and the process of information integration, an experiment was designed in which the individual's weighting of the importance of base—rate and diagnostic information was measured. In a first step, we will introduce a simple model that takes these information weights into account. When referring to the introduced variants of the Cab problem, we will demonstrate that weighting or reassessment of information in word problems is by no means unreasonable for some of the typical base—rate problems, and can yield solutions that deviate from the 'normative solution'. Finally, some empirical evidence for subjective reassessment will be reported.

The model: The model will be called the **individualized normative solution**. Its conception is simple. First, subjective parameters p'(H) and p'(D|H) are derived from the story parameters p(H), p(D|H), and the individual's information weights w(H) which is the weight given to the base—rates, and w(D|H) which is the weight given to the diagnostic information. The individualized normative solution is now the Bayesian solution p'(H|D) based on the subjective parameters. Let us assume that subjects' weightings are scaled between 0 and 1, then the subjective probabilities p'(H) and p'(D|H) are calculated as follows:

$$p'(H) = p(H) + (1 - w(H)) \cdot (.5 - p(H))$$
$$= w(H) \cdot p(H) + (1 - w(H)) \cdot 0.5$$
$$p'(D|H) = p(D|H) + (1 - w(D|H)) \cdot (.5 - p(D|H))$$
$$= w(D|H) \cdot p(D|H) + (1 - w(D|H)) \cdot 0.5$$

These equations convey that the smaller the weights w(H) or w(D|H), the more regressive toward the .5 probability the subjective probabilities p'(H) and p'(D|H). For example, knowing nothing about the base—rates or complete neglect of base—rates thus implies that H and $\overline{H}$ are modeled as being equally likely; analogously for the diagnostic information. The subjective probabilities p'(H) and p'(D|H) are taken for the calculation of the individualized normative solution for the normal version of the Cab problem and its variants:

$$p'(H|D) = \frac{p'(H) \cdot p'(D|H)}{p'(H) \cdot p'(D|H) + (1 - p'(H)) \cdot (1 - p'(D|H))}$$

Evidently, a linear interpolation between the given story parameters p(H) or p(D|H) and the .5 value via the subjective weights w(H) and w(D|H) is a rather rough approach. But we know that in many fields of decision research, linear approximations (DAWES, 1979) or weighted averages (cf. PITZ, 1980)

CHAPTER 2.4.

have shown a "robust beauty". No doubt one cannot expect that the inferential guideline of this model, which is the BAYES' theorem, describes the inferential process of all subjects. Nevertheless, the individualized normative solution starts from subjective parameters p'(H) and p'(D|H) which reflect the emphases given to the information. What these subjective parameters look like in the case of complete or partial neglect of base—rate and diagnosticity can be seen by studying the following model properties:

1. If w(H)=0 and w(D|H)= 0, then p'(H|D)=.5.
   According to the principle of insufficient reason, .5 will result if no information is taken into account.
2. If w(H)=0 and w(D|H)=1, then p'(H|D)=p'(D|H). No weight on the base-rate and full weight on the diagnosticity will yield a diagnosticity response p(D|H):
   Oppositely,
3. If w(H)=1 and w(D|H)=0, then p'(H|D)=p(H).
4. If w(H)=1 and w(D|H)=1, then p'(H|D)=p(H|D).
   The 'normative solution' will be predicted if both informations are given full weight, and:
5. If 0<w(H)<1 and 0<p(D|H)<1, then
   p(H)<p'(H|D)<p(D|H). For intermediate weights, values from the interval (p(H), p(D|H)) will result. Statement 5, however, is only always fulfilled for the normal version of the textbook paradigm.

Weighting in accordance with reasonable text interpretation: We will now supply illustrations for how an information weighting and the individualized normative solution may be linked to the reasonable reassessment of probabilities in the above examples (cf. Chapter 2.2.). In Example 2, for instance, the diagnosticity is already partially integrated into the base—rates, or, to express it in other words, the diagnosticity information already encompasses part of the base—rate information. This lowers the importance of the given diagnosticity information compared to the base—rate information. If one assumes that the weights w(H)=1 and w(D|H)=.91 are given, then the solution proposed in this example would result in p'(H|D)=.679. Usually there is a bunch of weights resulting in p'(H|D). For instance, w(H)=.86 and w(D|H)=.6 would also produce p'(H|D)=.679.

Illustration 2 in Example 3, Chapter 2.2. implies a lower significance for diagnosticity relative to base—rates, which were taken as truth and not subjectively altered. Weights w(H)=1 and w(D|H)=.51 yield p'(H)=.15 and

$p'(D|H) = .705$, and thus $p'(H|D) = .345$.

Empirical evidence for subjective reassessing of probabilities: Not only the proposed examples, but also the above—mentioned context features (representativeness, causality, availability, and specificity) have one thing in common, namely that the stories and their parameters are not assumed to be mapped one—to—one into the subject's mind. The individualized normative solution is a rough model which simply accounts for the possibility that subjects might start from different parameters than those given in the story. Surely, there may be many reasons that may lead to a reassessment of story parameters. Perhaps the examples and illustrations create a rather artificial impression, although they all have a concrete or factual history. Example 2 was created when tracing a subject's problem understanding. Example 1 refers to signal detection theory, Example 4, as far as we know, to a young doctor's question after listening to one of the illuminating lectures given by KAHNEMAN and TVERSKY, and Example 3 is based on the assumption that subjects might use their knowledge of the world and straightforwardly modify the story parameters.

2.5. SOME HYPOTHESES ON THE DEGREE OF BASE—RATE CONSIDERATION

Clearly, the above empirical evidence is no substitute for a critical test on the model's validity. Besides this model test, we will present several questions on the impact of various variables which are considered to affect behavior in base—rate problems. Though not all questions are derived from the above theoretical considerations, we will formulate them in the form of hypotheses in order to clarify the plan of the data analysis and to facilitate an understanding of the abundance of experimental results. We will first present the hypothesis on the individualized normative solution. If this model reflects essential features of the internal representation and the cognitive process, the deviation of the subjects' probability judgments from the individualized normative solution should be smaller than the deviation from the 'normative' solution.

Hypothesis 1: The subjects' response behavior on problems of the Cab type can be better predicted by the individualized normative solution than by the 'normative' solution which is merely calculated from the story parameters.

CHAPTER 2.5.

We shall already point out here that the information weights will be measured in a rather direct manner (by the subjects' ratings), and that there are different approaches to the conceptualization of weight (cf. SHANTEAU, 1980). Our procedure essentially differs from the commonly applied a posteriori procedure of weight estimation from the probability responses (cf. e.g., BIRNBAUM & MELLERS, 1983). Hence *Hypothesis 1* is by no means as trivial as one might suspect because it takes additional parameters into consideration. Of course, applying the a priori procedure of asking for the information weights might affect information processing. We could even expect that the questions on the importance or relevance of base−rate information might, for instance, cause this information to be more closely considered. Thus this potential effect will be controlled within our experimental design although it will not be used for the formulation of any hypotheses.

Subjects' response behavior in base−rate problems will be analyzed both with a 'normative' base line and from the model's (the individualized normative solution's) perspective. The next hypothesis deals with the variable **Age and educational** level which naturally is considered to be a crucial factor for the performance and the way in which problems are treated. This hypothesis and the following ones do not refer to the individualized normative solution but investigate the impact of certain salient task and performer dimensions.

Let us first look at the expectations that may be formulated with respect to the behavioral data if these are compared with the 'normative' solution. Clearly one critical question is: "Does ... formal training", which is gained in the course of education, "modify the untutored heuristics of everyday inductive reasoning?" (NISBETT, KRANTZ, JEPSON, & KUNDA, 1983, p. 340).

Traditionally, perhaps one should expect that increase in Age and level of education will lead to improved probability judgment. This inference supposes that there is a positive correlation between this variable and knowledge about probability calculus. Some evidence for this hypothesis may be gained from research into mathematics education (cf. GREEN, 1982a, b) or into developmental psychology (cf. PIAGET & INHELDER, 1951; HOEMANN & ROSS, 1982, pp. 111; SIEGLER, 1981). When using simple probability tasks such as chance perception in spinner or urn experiments, an increase in correct answers may often be observed in the course of adolescence. Hence one could presume that older, more highly educated subjects' (e.g., postgraduate students) responses are less 'biased'; for instance, show more correct answers and an increased incorporation of base—rates into their responses.

However, there are also some arguments stating that one may expect no improvement with increasing age, but perhaps even a deterioration of performance in some probability tasks. For instance, GREEN, 1982, revealed that subjects' performance in identifying typical random patterns of snow flakes declined with age. He concluded (p. 774) when interpreting this and various other results, "It can be hypothesized, that we see at work here two opposing tendencies — maturation/experience on the one hand and dominance of mathematical/scientific deductivism on the other which stifles the appreciation of randomness by seeking to codify and explain everything". Clearly in base—rate problems such a deterministic deductivism may lead to a cancelling out of relevant stochastic information, a phenomenon which has been modeled by the causal schema. However, there are also other findings that indicate a poorer performance by older samples compared with younger ones. In a study on adolescents' ability to combine probabilities, ROSS and DE GROOT, 1982, observed that in the age range from 12 to 19 no increase in the ability to determine disjunctive probabilites could be found. Here again, the reverse holds true, certain error types were significantly more often to be found among the 19—year—old (compared to 15year—old) subjects, if the probability of $p(A \cup B)$ had to be determined from $p(A)$ and $p(B)$, and $p(A \cap B)$ which had been implicitly provided by the task. The so—called "stand pat" choice, i.e., an unconsidered choice of some number which is displayed anywhere in a text, has been especially identified as a typical older student's error.

As there is a rather balanced pro and contra evidence for both an increase and a decrease in base—rate consideration, we will not formulate a one—sided hypothesis but propose:

Hypothesis 2: Base—rate neglect/consideration behavior differs with increasing level of age and education.

There are two more hypotheses that will be tested. The first is on the magnitude or extremity of base—rates and its impact on the response distribution.

Apart from the relational and contextual features, the numerical values are clearly the most important task variables in the Cab problem and its variants. There is some evidence and some points in favor of using and incorporating the extreme base−rates. TVERSKY and KAHNEMAN, 1979, p. 63, report that BAR−HILLEL's, 1975, unpublished doctoral dissertation documents some effects of extreme base−rates. The appearance or nature of these effects was not reviewed. There are some other studies in which the extremity of base−rates were manipulated. Both FISCHHOFF, SLOVIC, and LICHTENSTEIN, 1979, and BIRNBAUM and MELLERS, 1983, reported (modest) sensitivity to base−rate variation. Yet their experimental procedure of sequentially presenting one and the **same story** with different base−rates (in FISCHHOFF et al. ranging from .02 to .98!) is not only rather artificial, but also obviously tries to compel or provoke response changes. The third study is that of LYON and SLOVIC, 1976. They only reported descriptive statistics on the interquartile range which is 'normatively' worse for extreme base−rates. In summary, the empirical evidence is equivocal. One reason for this ambiguous empirical evidence on the impact of extreme base−rates might be found in the measures typically reported in research on the 'base−rate fallacy'. A measure reflecting the extent of subjects' information integration which allows for comparisons between different base−rate parameters has not been developed. This study attempts to fill this gap (cf. Chapter 2.6.5.).

One may also speculate about the differences between extreme and medium base−rate parameters. For instance, one may ask whether base−rates at the limits of the probability continuum are more salient or not (cf. NISBETT & ROSS, 1980), or whether perhaps a kind of Socratic dialogue will be induced such as: "One percent Blues is near to zero. What is the probability that the cab involved in the accident was Blue rather than Green if the base−rate for Blue is zero? 80%? This can't be true, as in that case there are no Blues. Hence ..."

With respect to the variation in base−rates, we will formulate the following hypothesis:

Hypothesis 3: The response behavior differs between the conditions of extreme and medium base−rates. Extreme base−rates are taken into account (relatively) more than medium base−rates.

The final hypothesis concerns learning through practice. It reads:

Hypothesis 4: In the case of iterated performance on different variants of the Cab problem, the significance of base−rate information will be increasingly realized, and base−rates will be increasingly incorporated into the judgment.

Once more, there are some findings and arguments that support this hypothesis and some against it. Let us first deal with the proarguments. On a superficial level, subjects in psychological experiments have to adjust to the artificial conditions under which they have to perform. This process of adjustment may be accompanied by stress reduction followed by step−by−step decentration (cf. LINDSAY & NORMAN, 1977; MANDLER, 1975), penetration into the problem structure, and detection of previous hidden variables (e.g., base−rates) or calculation errors (cf. ALLWOOD & MONTGOMERY, 1982). BIRNBAUM and MELLERS, 1983, found extreme changes between the first judgment and a later judgment on one and the same problem which had been embedded among 100 consecutive trials. However, the response distribution they report on the embedded consecutive response does not significantly differ from a normal distribution with mean .48 and a standard deviation of .14 (a KOLMOGOROV−SMIRNOV test provides $p < .35$ calculated according to the data presented in BIRNBAUM & MELLERS, 1983), and there is not even a significant difference to a normal distribution with mean .5 and the above standard deviation. This randomization of response behavior is not surprising, especially if one takes into account that the 101 responses in the BIRNBAUM and MELLERS, 1983, experiment were given by undergraduates within less than two hours! Admittedly, on a closer analysis, there may be some counterevidence to a totally normally distributed random response with a mean of .5, but there are also further essential procedural differences between BIRNBAUM and MELLERS' and our experiment. As far as we know, no systematic study has been made on whether experience from one base−rate problem will result in a positive transfer to another. This will be investigated in our experiment. However, two conditions must be noted which may lessen this learning through repetition: First, no feedback will be given about the responses after each trial, and second, according to experiences reported by mathematics educators (cf. DINGES, 1979; FREUDENTHAL, 1973; v. HARTEN & STEINBRING, 1984), the difficulties of the subject matter and the barriers that hinder the development of stochastic thinking are scarcely to be overcome within short run interactions.

The following Experiment A was designed to test the above hypotheses. Various subject samples with different age and educational levels (*Hypothesis 2*) dealt with a series of base–rate problems. In this experiment subjects were asked to rate the significance of the base–rate and diagnosticity information. Hence the experimental data allow for a test of the model's (i.e., the individualized normative solution) validity (*Hypothesis 1*). Furthermore, the base–rate parameter was systematically manipulated to test *Hypothesis 3*. As each subject worked on a series of problems, the effect of practice (*Hypothesis 4*) could also be controlled.

2.6. EXPERIMENT A

2.6.1. SUBJECTS

The subjects were 137 student volunteers from three different grade levels of a German Gesamtschule (comprehensive high school) and 43 postgraduate students of Bielefeld University who had received their Vordiplom (a pre–diploma, received after two years of university study). Three of the university students did not complete the task and were omitted from the data analyses.

2.6.2. PROCEDURE AND EXPERIMENTAL TASKS

Subjects participated voluntarily in a 'questionnaire study on information processing' in groups or individually, and received DM 5.00 for their participation. In the beginning, they were told that complete concentration and attention were indispensible in order to survey the valid data. The experimenter always asked the subjects to interrupt their work if they felt they were losing their concentration or interest. There was a time limit of one hour. Subjects were asked to get involved and were instructed to work on the problems as if they had been put to them by a close friend for whom they were important and actually relevant. Paper and pencil were available.

The questionnaire included the **Hit Parade**, the **Motor**, the **TV** problem, and one non–Bayesian problem which is omitted in this volume. The texts of these problems are given in Chapter 2.1.

Half of the questionnaires additionally contained questions on the information weights of the base−rate information and the diagnosticity information for all three base−rate problems. The subjects were instructed to deal with the problems sequentially. They were not allowed to leaf back through the pages of the questionnaire or to correct prior responses. The questions on the information weights had to be answered before the probability judgment was given. For the Hit Parade problem, for example, these were:

Base−rate question: Is the information that 35% of the songs newly presented became hits of significance for the probability to be assessed?

Diagnosticity question: Is the studio guest's 80% correct forecast for both hits and non−hits of significance for the probability to be be assessed?

The responses were given on a 6−point scale with the poles denoted "not at all" and "very significant". The probability judgment responses had to be given as percentages.

About half of the subjects also received a postexperimental questionnaire asking for ratings of their involvement and interest in the tasks. They also had to rate each story for realism and the attractiveness of its content.

2.6.3. INDEPENDENT VARIABLES AND EXPERIMENTAL DESIGN

The experimental design was a 4 x 2 x 2 x 4 factorial design with factors Age, Base−Rates, Information Weights, and Order. As far as possible, the cell frequencies were uniformly distributed.

Age and educational level had four levels:
Age 1:   39 seventh grade students ($\bar{x}=13.3$, $s=0.68$),
Age 2:   47 ninth or tenth grade students ($\bar{x}=16.0$, $s=0.91$),
Age 3:   51 twelfth grade students ($\bar{x}=18.5.$, $s=0.83$), and
Age 4:   43 postgraduate students ($\bar{x}=24.3$, $s=2.6$).

Extremity of base−rates: About half of the subjects were given extreme and the other half medium base−rates. The diagnosticity was kept constant for each story (see below, Table 2.1). For both levels, i.e., for both medium and extreme base−rates, the parameter was slightly varied between the stories. This was done in order to prevent a recognition of the structural equivalence of the different base−rate problems through a comparison of the numerical values, and also to prevent a sterotyped response behavior. The base−rate parameters actually chosen were; for the medium base−rate level: Hit Parade

.35, TV .25, Motor .30; and for the extreme base–rate level: Hit Parade .10, TV .02, Motor .05. The **'normative solutions'** by medium base–rates were for the Hit Parade .68, the TV problem .75, and the Motor .63. By extreme base–rates the 'normative solution' yields for the Hit Parade problem .31, for the TV problem .16, and for the Motor problem .17.

Order: The three base–rate problems and the non–Bayesian problem labeled NB were presented in the following orders: Motor, Hit, NB, TV; TV, Motor, Hit, NB; NB, TV, Motor, Hit; and Hit, NB, TV, Motor.

Information weights were a two level factor. Half of the questionnaires contained questions about the information weights, the other half not. The two levels were introduced in order to control the potential effect of inquiring into the weights (cf. FISCHHOFF & BAR–HILLEL, 1984b).

2.6.4. DEPENDENT VARIABLES AND MEASURES

Before turning to the results section, we wish to discuss the measures which reflect a base–rate neglect/consideration. If we look at the data from the reports on the Cab problem (TVERSKY & KAHNEMAN, 1979; BAR–HILLEL, 1980; WELL, POLLATSEK, & KONOLD, 1982), we find that only the percentage of responses identical with the additional information is discussed and that the medians are compared. From a measurement theory perspective, these are weak strategies which only permit rough analysis.

In addition to the subjects' probability responses and their ratings of information weights, three measures on the adequacy of subjects' performances were used. The first, d_1, is the signed difference between the subject's response and the 'normative solution', $r - p(H/D)$. However, this measure is not suitable for comparisons across problems with differing parameters. Therefore a new measure, d_2, was developed specifically to make such cross–problem comparisons more meaningful. The measure d_2 has the following characteristics:
1. When $r = 0$, $d_2 = -.5$
2. When $r = 1$, $d_2 = .5$
3. When $r = p(H)$, $d_2 = -.25$
4. When $r = p(D|H)$, $d_2 = .25$
5. When $r = p(H|D)$, $d_2 = 0$

The measure d_2 is calculated as follows:

$$d_2 = \begin{cases} \dfrac{.25 \cdot r}{p(H)} - .5 & \text{if } r \leq p(H) \\[6pt] \dfrac{.25 \cdot (r - p(H))}{p(H|D) - p(H)} - .25 & \text{if } p(H) < r \leq p(H|D) \\[6pt] \dfrac{.25 \cdot (r - p(H|D))}{p(D|H) - p(H|D)} & \text{if } p(H|D) < r \leq p(D|H) \\[6pt] \dfrac{.25 \cdot (r - p(D|H))}{1 - p(D|H)} + .25 & \text{if } r > p(D|H) \end{cases}$$

The properties of the measures d_1 and d_2 are illustrated for different values of $p(H)$ in Figure 2.3. In both figures the upper lines present response scales for medium, the lower lines response scales for extreme base−rates. The parameters for the base−rates, the 'normative solution', and the diagnosticity information are inserted. In Figure 2.3.a the middle line displays the error measure d_1, analogously the middle line in 2.3.b presents the error measure d_2. The connecting lines between the upper (lower) and the middle line indicate how the responses 0, $p(H)$, $p(H|D)$, $p(D|H)$, and 1 are mapped onto the d_1 or the d_2 scale.

As can be seen from Figure 2.3.a, for d_1, a base−rate response, $p(H)$, yields a much higher deviation from zero than a diagnosticity response, $p(D|H)$, by medium base−rates, whereas the opposite is true for extreme base−rates. If the absolute value of d_1 is taken as a yardstick for the deviation in the TV problem with extreme base−rates, a complete neglect of base−rates (i.e., a diagnosticity response) would produce a deviation which is 5.28 times as large as a complete neglect of diagnosticity information (i.e., a base−rate response). However, if the absolute value of d_2 is applied, a base−rate response and a diagnosticity response are judged as equally fallacious (cf. Figure 2.3.b), and a comparison of responses given on problems with medium and extreme base−rates also seems to be justified. For intermediate responses, that means, responses which lie in one of the intervals (0, $p(H)$), ($p(H)$, $p(H|D)$), ($p(H|D)$, $p(D|H)$), and ($p(D|H)$, 1), the d_2 measure is yielded by linear interpolation between the d_2 values of the interval limits.

A third measure, d_3, is similar to d_2 except that the calculations are derived from the subjective problem parameters estimated from the rescaled importance weight ratings, that is, replacing $p(H|D)$ by $p'(H|D)$ within the above calculations.

Figure 2.3.: Illustration of the properties of the measures d_1 and d_2 for probability estimations given on base–rate problems with different values for $p(D|H)$ and $p(H)$ (The values are taken from the TV problem)

a) Illustration of the properties of d_1:

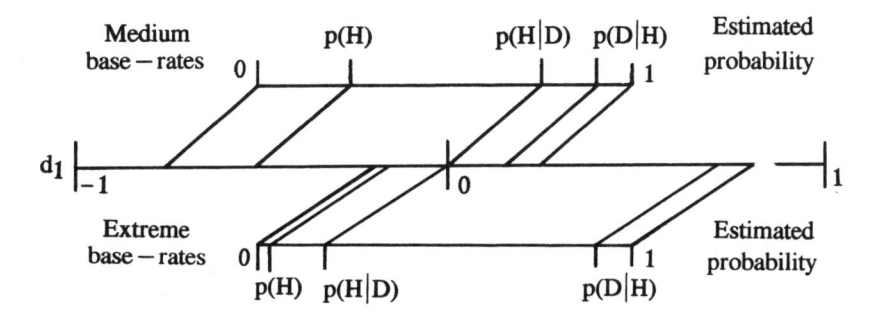

b) Illustration of the properties of d_2:

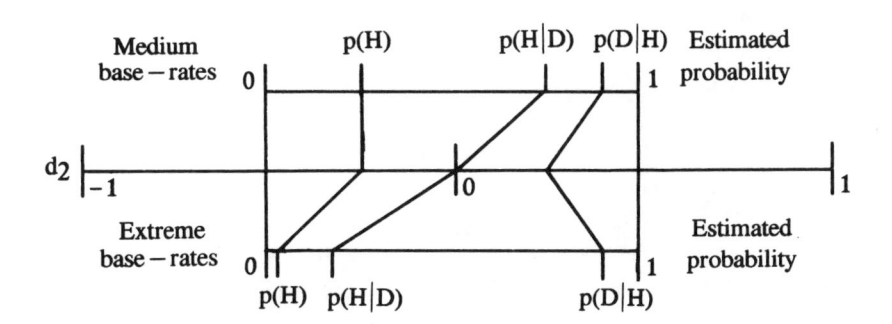

In the analyses, priority is given to the absolute values of the measures d_2 and d_3 so that the errors of responses above and below the 'normative solution' do not cancel each other out when the means are considered.

2.6.5. RESULTS

Preliminary remarks on the presentation of the results: Data and analysis of data will be presented in a rather detailed and exhaustive manner that includes six tables and eight figures. This is done in order to provide insight into the consistency and structure of the results, and to enable the reader to check hypotheses, for instance, about the differential impact of the base − rate stories, that are not treated in depth in this chapter.

After first reporting some data from the **postexperimental questionnaire**, histograms on the **response distribution** and statistics for the mean deviations, d_1, will be presented.

The experimental plan permits the use of an analysis of variance. Although, according to theoretical considerations, the conditions for a parametric analysis presumably might not be fulfilled or may not be controlled due to low cell frequencies, because of the nonavailability of nonparametric analysis methods in multivariate − analysis (cf. PURI & SEN, 1971; LEHMANN, 1975), an **overall analysis of variance** will be performed on $|d_1|$ and $|d_2|$ in order to grossly inform us about the strong effects of the independent variables and interactions.

The **hypotheses** themselves will then be tested principally by one − factorial nonparametric methods, while on occasion, various tests (e.g., a KOLMOGOROV − SMIRNOV and a MANN − WHITNEY U − test) will be simultaneously conducted to elucidate different aspects of the response behavior, and to demonstrate the presence or lack of robustness of effects. In this context, parametric tests will also be supplementarily provided. Furthermore, we want to note that an adjustment of the significance level (e.g., by BONFERRONI's or other procedures like the HOLM or RÜGER procedure, cf. ABT, 1983), which should be performed because of the iterated univariate testing of one and the same data, will not be applied as neither the number of tests actually calculated have been determined in advance, nor are the correlations between the test variables known to prevent ultraconservative testing. Due to the missing alpha − adjustment some of the results have to be interpreted with reservations. The presentation of results will start with effects of base−rates (*Hypothesis 3*), this is then followed by age effects (*Hypothesis 2*), and the analysis of the impact of iteration (*Hypothesis 4*) and information weights. At the end of the results section, the individualized normative solution will be tested (*Hypothesis 1*).

It should be noted that all results are reported that have significant ($p<.05$) or highly significant ($p<.01$) effects. However, the level of marginally significant results ($p<.10$) will also be reported in order to provide information about the strength of trends.

Postexperimental questionnaire: The postexperimental questionnaire indicates that the subjects got very involved and judged the problems to be of medium interest. There is a median of 5 for involvement and of 4 for interest, both on 6−point scales. The results are presented by box−and−whisker plots (cf. TUKEY, 1977; BIEHLER, 1982) which are appropriate displays for discrete rating scales. These plots (cf. Figure 2.4) visualize the loci of the median relative to the first and third quartile. The ratings of realism on 7−point scales were highest for the Hit Parade with a median of 6, somewhat lower for the TV, and lowest for the Motor problem, both with a median of 5.

Description of response distributions: The histograms in Figure 2.5 provide a sufficient statistic of subjects' response distributions. We have split the distributions according to base−rates for each story. This is done because the response distributions are strongly determined by the base−rates, which would be concealed if both distributions were mapped as one. All distributions are bimodal with an upper peak on the diagnosticity and a lower peak on or near to the base−rate. The percentages of base−rate responses vary between the stories. Nevertheless, if all problems and base−rates are pooled, only slightly more than 20% of the subjects respond with the diagnosticity. The necessity of splitting the response distributions according to base−rates is also substantiated by the d_1 measure. When medium (low) base−rates are considered, the group means indicate underestimation for all problems, and when extreme (low) base−rates are considered, there is overestimation. The difference is considerable as can be seen from the d_1 values in Table 2.1. A U−test indicates (marginally) significant differences: $p<.001$ in the Motor; $p<.07$ in the Hit; and $p<.02$ in the TV problem.

Overall analysis of variance for $|d_1|$ and $|d_2|$: As already mentioned above, the hypotheses will predominantly be tested with the absolute value of

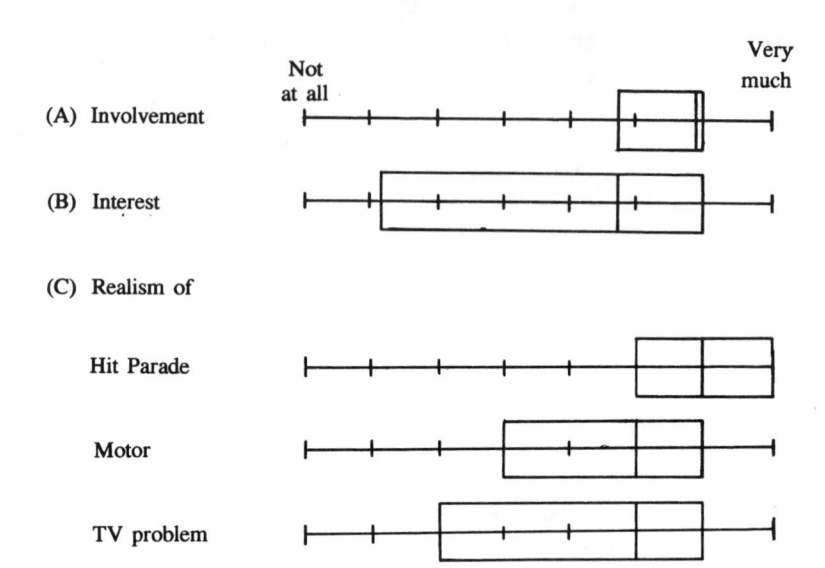

Figure 2.4.: Box–and–whisker plots for the question on the subjects' involvement (A), interest (B), and rating of realism (C) for each story

Table 2.1.: Means of d_1, $|d_1|$, and $|d_2|$ for all base–rate problems split according to base–rates

Problem	Motor		Hit		TV	
Base-rates p(H)	.30	.05	.35	.10	.25	.02
Diagnosticity p(D|H)	.80	.80	.80	.80	.90	.90
'Normative solution' p(H|D)	.63	.17	.68	.31	.75	.16
Mean d_1	-10	22	-14	15	-36	18
Mean $|d_1|$	24	29	24	30	42	27
Mean $|d_2|$	23	18	24	21	27	17

d_1 and d_2. Although d_1 is bimodally and evidently nonnormally distributed, this cannot be stated for its absolute value $|d_1|$ or for $|d_2|$. Although

CHAPTER 2.6.5.   43

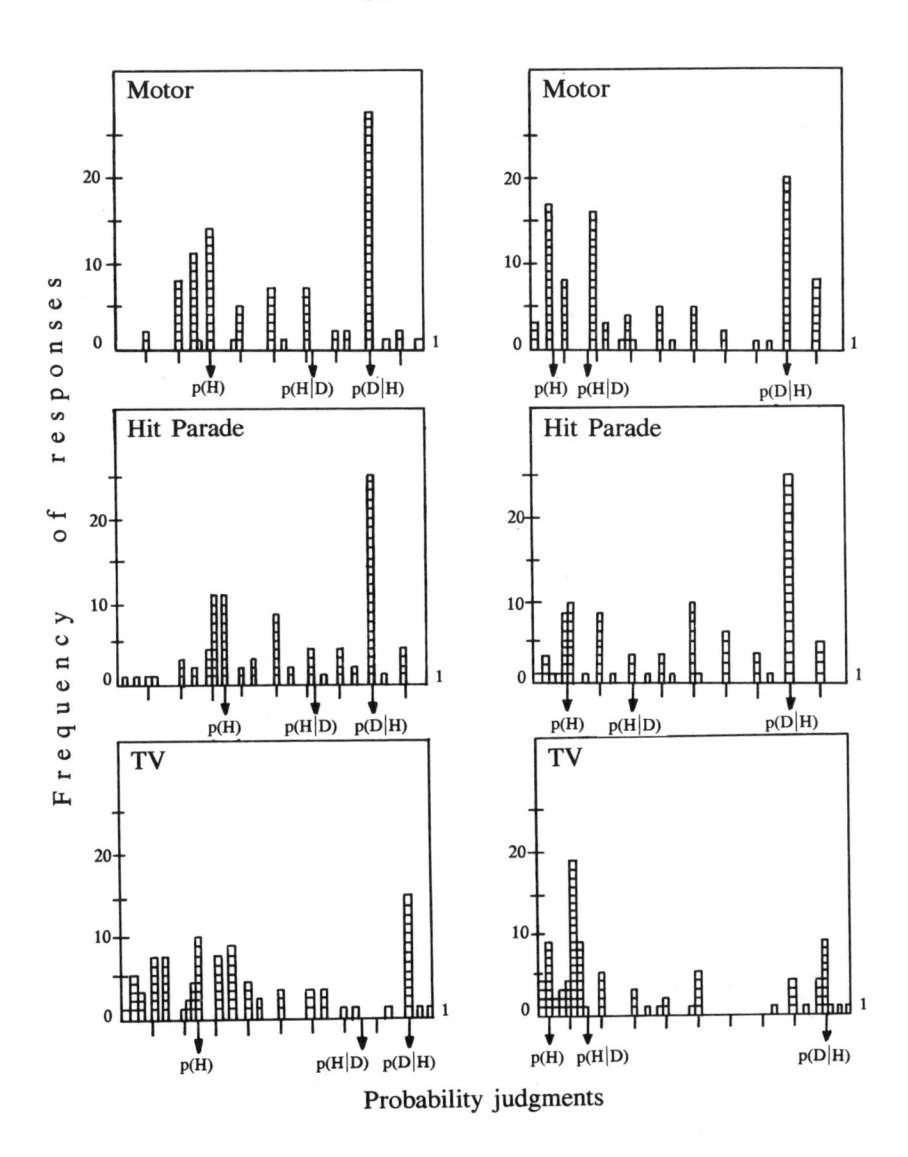

Figure 2.5.: Histograms of response frequencies separated for medium (left side graphs) and extreme (right side graphs) base − rates and separated for the different stories (X − axis: estimated probability; Y − axis: frequency)

proponents of the analysis of variance often dismiss the need for a test of the assumptions (i.e., normal distribution for each cell, homogeneity of variances for each treatment combination, independence of errors), by referring to BOX's, 1953, 1954, fundamental papers on the robustness of the F−test with respect to violations of these assumptions, we will take a closer look. Due to the low cell frequency (on average about three), a test on normality and homogeneity of variance should not be performed if all factors i.e., Base−Rates, Age, Information Weights and Order are simultaneously integrated into one design. Therefore testing was performed with a reduced design with the factors Base−Rate and Age. A KOLMOGOROV−SMIRNOV test yielded one significant result for $|d_1|$ in each of the eight tests on each problem. For $|d_2|$, two of the tests on the Hit problem, three of those on the Motor, and one of the eight tests on the TV problem were significant. Hence there are six significant results compared to 1.2 which should be expected on average in these 24 tests at a .05 level of significance.

The findings from the BARTLETT test on homogeneity of variance are worse. Here half of the tests are significant. The $|d_1|$ value for the Hit problem and the $|d_2|$ values for the Motor and TV problems are significant, indicating different variances in the cells. The statistical independence among the error components was tested by PEARSON's Chi^2 for the reduced design with the Base−Rate and the Age variables. According to this test independence may be assumed.

Anyhow, the results of the analysis of variance have been interpreted with caution and should be considered as more or less descriptive statistics. According to the analysis of variance, the base−rates are the most crucial variable. For all problems, both $|d_1|$ and $|d_2|$ differ with at least marginal significance (cf. Table 2.2.). The effects of base−rates may be studied in Table 2.1. The Motor and Hit problems show a significantly lower mean for $|d_1|$ with medium base−rates. As Figure 2.5 (cf. also Figure 2.3) reveals, the high $|d_1|$ value for medium base−rates is due to the large weight of the diagnosticity responses, for instance, compared to the base−rate responses. This is an unwanted property, as a complete neglect of one of the two given informations, either the base-rate information or the diagnosticity information, should not be given preference by the measures applied. Thus the $|d_2|$ measure will be used in the subsequent analyses.

For the sake of completeness, we will also present a graphical representation of the significant interaction of Age and Base−Rates for $|d_1|$ (cf. Figure

Table 2.2.: Significant and marginally significant main effects and first order interaction of analysis of variance, design : Age x Base−Rate (BR) x Information Weights (IW) x Order (O), for $|d_1|$ and $|d_2|$

Dependent variable	Source	MS	df	F	p <
Motor problem					
$|d_1|$	BR	972.2	1	3.14	0.07
	Error	310.1	112		
$|d_2|$	BR	966.1	1	8.02	0.01
	Error	120.5	112		
Hit Parade problem					
$|d_1|$	BR	924.0	1	3.76	0.06
	BRxAge	772.4	3	3.14	0.03
	Error	246.0	112		
$|d_2|$	BR	422.3	1	3.75	0.06
	Error	112.7			
TV problem					
$|d_1|$	BR	7078.4	1	12.58	0.001
	IW	2172.1	1	3.86	0.06
	BRxAge	1713.61	3	3.04	0.04
	Error	562.8	113		
$|d_2|$	BR	3872.4	1	32.9	0.001
	Error	117.8	113		

2.6). These results will be discussed in detail after the section on age effects. We will only mention here that the interactions may be explained by the differential weighting of base−rates *(Hypothesis 3)*.

Through all problems, the $|d_2|$ values are consistently and at least marginally significantly different for the two base−rates.

Base−Rates *(Hypothesis 3):* As reported above, the manipulation of base−rates clearly affects the response distributions and the form of this effect can be inferred from Figure 2.5. The distributions are pulled in the direction of the bottom of the probability scale by extreme (low) base−rates. The nonparametric U−test indicates highly significant lower probability judgments for extreme base−rates compared to medium base−rates in the Motor

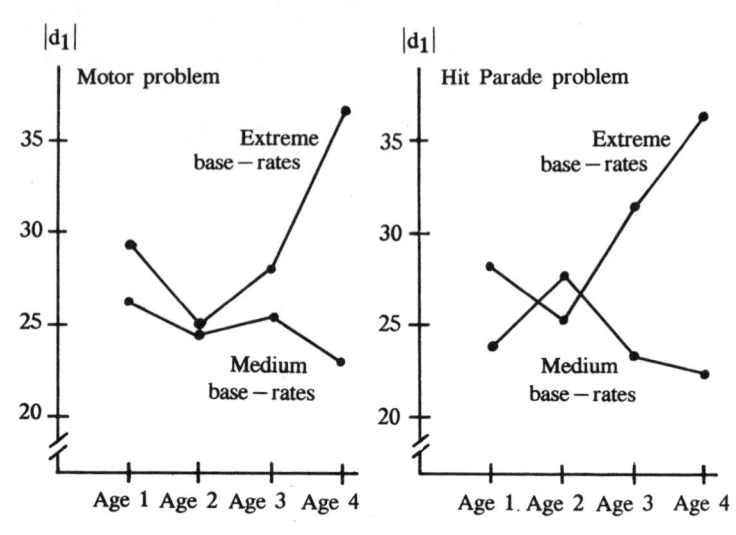

Figure 2.6.: Graphical representation of the interaction of Age and Base−Rates for $|d_1|$ in the Motor and Hit Parade problems

problem, and significant lower judgments in the Hit and TV problems.

In some respects, the $|d_2|$ measure reflects the degree of base−rate consideration. Here again, the U−test confirms the analysis of variance in that it produces significance levels of $p<.001$ for the Motor, $p<.07$ for the Hit, and $p<.001$ for the TV problem. A weak indicator for the degree of base−rate consideration is the frequency of diagnosticity responses, which is smaller for the extreme Base−Rates in the Motor and the TV, yet about the same for the Hit problem. The difference for the Motor problem is significant, $p<.05$. However the second part of *Hypothesis 3*, which concerns the differential impact of extreme and medium base−rates, could not (fully) be confirmed, as there is no clear pattern, and the distributions do not indicate that extreme base−rates are definitely taken more into account.

Age *(Hypothesis 2):* Information about the **response distributions** split by age is displayed in Table 2.3. Response behavior is clearly affected by the age and educational level. A Chi^2 analysis for the classification presented in Table 2.3. yields significant differences for the Motor ($Chi^2=28.3$, df=12, $p<0.01$), the Hit ($Chi^2=29.73$, df=12, $p<0.01$), and also the TV problem ($Chi^2=17.8$, df=9, $p<0.05$). There are fewer degrees of freedom in the TV problem as the responses greater than the diagnosticity were tied to the diagnosticity responses, due to inadmissably low cell frequences.

Some more information about the sources of the differences in response behavior and the different degrees of information integration by the various age groups may be gained when analysing the single response categories of Table 2.3. So, for instance, the younger age groups show considerably more **middling responses**, i.e., responses in the (p(H), p(D|H)) interval. This trend is consistent across all problems. We will also report some Chi^2 statistics on this, as they provide better information about the strength of differences than can be gained from just studying percentages. We want to stress again, however, that these statistics have to be treated as descriptive statistics, which do not permit inferential conclusions. Furthermore, one has to take into account that the subsequent Chi^2 – values are statistically dependent and thus provide redundant information. We will now look at these middling responses. The Chi^2 – values indicate 'significance' for the middling responses in all three problems; for the Motor (Chi^2=9.05, df=3, p<.05), the Hit problem (Chi^2=15.5, df=3, p<0.01), and the TV problem (Chi^2=9.4, df=3, p<.05).

Younger subjects are much less prone to produce **diagnosticity responses**. Once again, this effect is consistent across all problems, and the Chi^2 statistics for the Motor problem are: Chi^2=9.7, df=3, p<.05; for the Hit Parade problem: Chi^2=18.55, df=3, p<.01; and for the TV problem: Chi^2=8.07, df = 3, p < .05. This shows us what the "other side of the coin", i.e., the 'middling responses', look like.

Finally, it should be noted that the younger age groups tend to produce fewer 'extreme errors', i.e., responses below the base–rates and above the diagnosticity, and also that the .5 value occurs with almost equal frequency in all age groups.

Although the mean $|d_2|$ is, on average, slightly larger for the two older age groups (cf. Table 2.4, Figure 2.7), this trend is not statistically significant for any of the problems. Obviously, the interaction of the Age and Base–Rate variables on the $|d_1|$ measure (cf. Figure 2.7) is caused by the low value of $|d_1|$ for a diagnosticity response in the case of medium compared to extreme base–rates (cf. Figure 2.2). The interaction of Age and Base–Rates thus merely seems to be an artefact caused by the high percentage of diagnosticity responses in the postgraduate sample (Age 4).

Table 2.3.: Effect of Age on response distributions (An attached 'm' indicates that the specific value is the mode of the response distribution. The term 'mid' denotes the open interval $(p(H), p(D|H))$)

Percentages of responses in range

Age groups	<p(H)	p(H)	mid	p(D|H)	>p(D|H)	.5
Motor problem						
Age 1 ($\bar{x} = 13.5$)	6	31m	36	17	11	6
Age 2 ($\bar{x} = 16.0$)	15	11m	54	11m	9	9
Age 3 ($\bar{x} = 18.5$)	22	10	35	29m	4	10
Age 4 ($\bar{x} = 24.3$)	12	12	33	40m	5	2
Hit Parade						
Age 1	19	6	50	17m	8	11
Age 2	22	11m	67	9	7	9
Age 3	24	8	27	37m	4	14
Age 4	19	16	19	44m	5	2
TV problem						
Age 1	19	3	68	11	0	16m
Age 2	26	11m	61	4	0	2
Age 3	16	17	43	18m	6	0
Age 4	19	5	51	19m	7	2
All problems						
Age 1	15	13	51	15m	6	11
Age 2	14	11	56	7	5	9
Age 3	20	12	35	28m	5	12
Age 4	16	11	34	34m	5	3
All Ages	18	12	44	21	5	7

Table 2.4.: Effect of Age on $|d_2|$

$|d_2|$	Age 1	Age 2	Age 3	Age 4
Motor problem	20.7	19.7	22.1	20.4
Hit parade	20.4	21.4	23.7	24.2
TV	18.8	21.1	23.1	23.1
All problems pooled	19.9	20.7	23.0	22.6

Iteration *(Hypothesis 4):* Hypothesis 4 concerns learning and the improvement of probability judgments during the course of repeated performance on similar problems. Table 2.5. shows the means of $|d_2|$ for the four positions at which one and the same problem is handled. If all problems are pooled, the third and fourth iteration show slightly smaller $|d_2|$ means. However, both

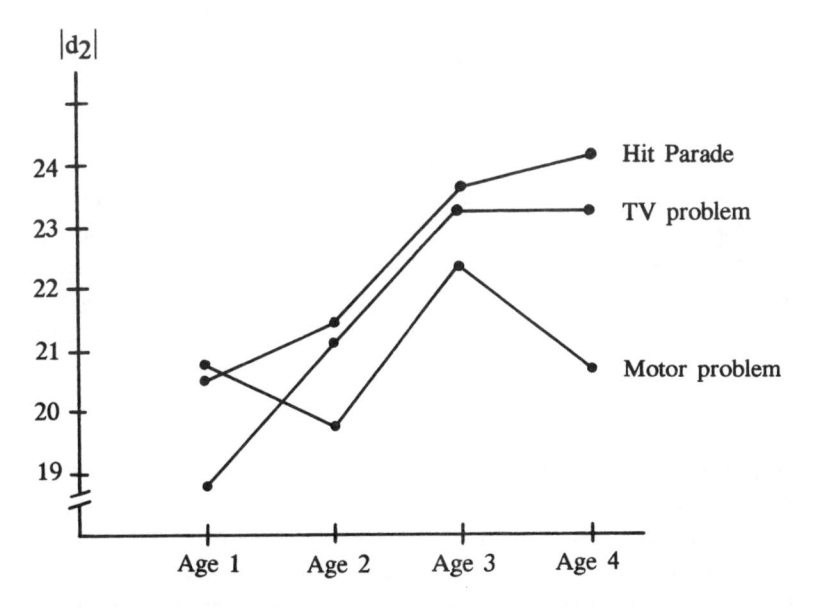

Figure 2.7.: Effect of Age on $|d_2|$

parametric and nonparametric analyses yield no significant results. We could find no significant effect for $|d_2|$ and $|d_1|$, or for the response distributions or relative frequencies of diagnosticity and base−rate responses.

Table 2.5.: Means of $|d_2|$ for the four positions P1−P4 at which the problems were introduced to the subjects

$|d_2|$	P1	P2	P3	P4
Motor	20.9	21.3	22.0	19.7
Hit Parade	20.5	22.9	22.4	21.9
TV	23.8	22.0	18.0	18.4
All problems pooled	21.8	22.0	20.8	20.1

Information weights and the individualized normative solution *(Hypothesis 1):* The experimental procedure of measuring the information weights did not significantly affect the response distribution. Apart from a marginally significant affect on the $|d_1|$ measure in the TV problem, no further test indicated any impact of the procedure of measuring information weights on the response behavior.

Before testing *Hypothesis 1*, we shall report on the distribution of information weights. It is clear (cf. Table 2.6.) that the significance ratings for the base−rates have a U−shaped distribution for all problems. The ratings on the importance of the diagnosticity information also have two peaks, although the distribution of the Motor problem more closely matches a skewed single peaked function type than a U−shaped one. Diagnosticity was given higher information weights. A sign test results in $Chi^2 = 4.04$, df=1, $p<.05$, if the sum score of the ratings on the three problems are submitted to the test procedure.

According to the principles of the individualized normative model's construction, a high probability judgment should be given by subjects who produce high diagnosticity and low base−rates weights, and low responses by those producing relatively low diagnosticity and high base−rate ratings.

In accordance with our expectations, there are significant correlations between the ratings of information weights and the performance. The analysis was performed on the d_2 measure as the base−rates are pooled. KENDALL's

Table 2.6.: Frequencies of ratings of importance of base−rates and diagnosticity

Rating

	1	2	3	4	5	6	n

Base-Rates

Problem

Motor	14	5	7	12	27	25	90
Hit Parade	22	13	12	10	10	24	91
TV	19	10	7	10	16	30	92
All	55	28	26	32	53	79	

Diagnosticity

Motor	5	4	10	12	17	41	89
Hit Parade	11	6	7	13	22	31	90
TV	6	2	10	12	15	47	92
All	22	12	27	37	54	119	

tau yields for the base−rate ratings, for the Motor problem: tau = −.38, $p < .01$, n = 90; for the Hit Parade: tau = −.43, $p < .01$, n = 90; and for the TV problem: tau = −.13, $p < .10$, n = 90. The correlation between the diagnosticity ratings and the d_2 measure is less extreme and yields for the Motor problem: tau = .19, $p < .05$, n = 89; the Hit Parade: tau = .23, $p < .05$, n = 89; and the TV problem: tau = .10, $p < .10$, n = 89.

We will now turn to the model of an individualized normative solution which may be calculated for each of the subjects who produced information weights if these are rescaled between zero and one. Another view of the relation of information weights and performance on information integration is provided by the comparison of the $|d_2|$ and $|d_3|$ measures. When all problems and subjects are pooled, there is no significant difference, and $|d_3| = 23.6$, based on the individualized normative solution, is worse than $|d_2| = 21.2$. If we split the subjects according to age groups, however, there is an interesting differential effect (Figure 2.8.). For Age 1 or 2, the mean differences between 'normative' solution and responses $|d_2|$ are larger than the differences between the model's prediction and the responses $|d_3|$. For Age 3, the means are about the same, but for the postgraduate students the mean of $|d_3|$ is considerably

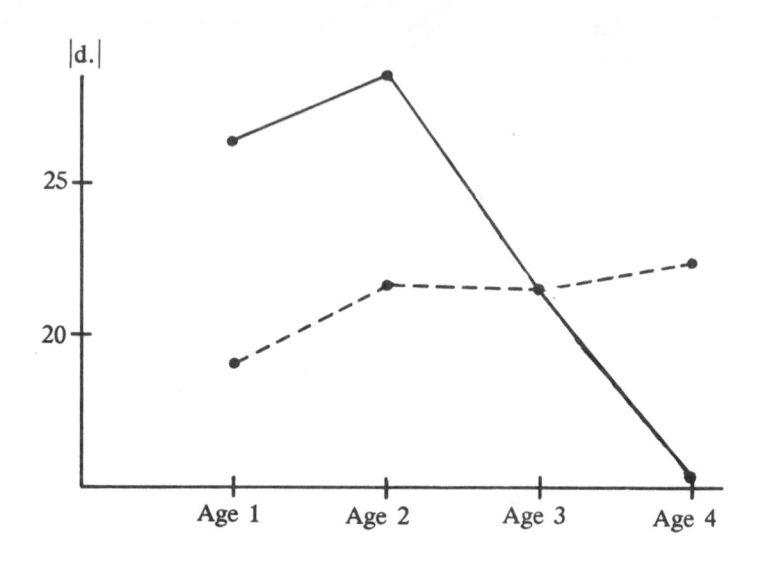

Figure 2.8.: Means of $|d_2|$ (---) and $|d_3|$ (———) separated for the age groups Age 1 to Age 4

smaller than $|d_2|$. An analysis of variance on the differences between $|d_3|$ and $|d_2|$ indicates a significant age effect, $F(3,88) = 5.98$, $p < .001$, which may also be confirmed by a more adequate nonparametric KRUSKAL–WALLIS analysis of variance ($p < .05$).

2.7. DISCUSSION OF EXPERIMENT A'S RESULTS

When looking for the essence of these findings, one should clarify which "situational generalizability" and which "theoretical psychological interpretations are supported" (cf. SCHOLZ, 1980a) by these results. Laboratory research into judgmental processes, heuristics, and biases has been criticized because of its neglect of task and performer dimensions (cf. EDWARDS, 1983; CHRISTENSEN–SZALANSKI & BEACH, 1982; COHEN, 1979). Furthermore, it has been stated that people are unmotivated, and that the "studies are grossly unrepresentative both of tasks and of subjects who might perform these tasks" (EDWARDS, 1983, p. 503). In the control given by the postexperimental questionnaire reviews, the present subjects showed high involvement and judged the experiment to be of interest. The Hit Parade problem taken from the young people's world appears to be highly realistic to them. We do not consider that these ratings reflect a large social desirability distortion, as the subjects gave low ratings on the attractiveness of the stories, especially by the Motor and TV problems. The following results deal with our studies on some crucial performer and task variables (cf. CALDWELL & GOLDIN, 1979; SAHU, 1983; EDWARDS, 1983) that seem to be of broader interest.

People are sensitive to base−rate variations: Response distributions on Cab problems are affected by the base−rate parameter. This is not only true within procedures of subjective sensitivity analysis (i.e., putting one and the same problem with different parameters to one and the same person; cf. BIRNBAUM & MELLERS, 1983; FISCHHOFF, SLOVIC, & LICHTENSTEIN, 1979), but also within the procedure of comparisons between subjects. TVERSKY and KAHNEMAN's, 1979, hypotheses that the observed probability judgments are relatively worse for extreme base−rates when compared to moderate base−rates, could not be confirmed. But although there is some evidence for the contrary as stated in *Hypothesis 3*, our results do not clearly support an opposite hypothesis. Our data cannot answer the question **why** the extremity of base−rates affects behavior. To answer this question, other methods have to be applied (see Chapters 3 and 4), but obviously response behavior is sensitive to the variation of numerical parameters.

At this point, we wish to mention that comparisons with BAR−HILLEL's data, 1980 (cf. Figure 2.2., 2.5.), collected from a small Israeli sample of students by a university entrance test, shows that one and the same problem story, i.e., the Motor problem, may elicit considerably different response patterns. Thus BAR−HILLEL's confirmatory results on the validity of the specificity construct could not be replicated by our data.

Iterated performance at similar problems is not enough: A short run iterated working at similar (but different) problems without feedback did not change the response distributions (cf. *Hypothesis 4)*. This result is in line with recent findings on the impact of instructions and tutorials on base−rate problems (cf. LICHTENSTEIN, 1984; FISCHHOFF & BAR−HILLEL, 1984b). Obviously, a spontaneous change of response behavior or a self−

organized learning can not be expected under the given experimental conditions.

Age and education do not improve performance: *Hypothesis 2* on the differential consideration/neglect of base−rates has been confirmed. Though there is no significant difference in the response distributions across all three different variants of the Cab problem, there is a considerably higher number of subjects who completely neglect base−rates within the older age groups. Hence, in some respects, older subjects are more biased than younger ones.

Other parameters also indicate rather greater deviations from the 'normative solution' in older rather than in younger subjects. Thus this finding is in line with a well−known phenomenon in developmental psychology, i.e., that increased age and increased education in some domains of behavior is accompanied by a change for the worse in performance (cf. PIAGET & INHELDER, 1951/1975; KARMILOFF−SMITH, 1982).

The developmental and educational level of our older subjects (and their response distributions) are similar to those of most other studies on the 'base−rate fallacy'. The critical question raised by these findings is: Why are older subjects more biased than younger subjects? A little insight into the answer may be gained by looking at the results on the information weights.

Information weights supply some insight into the judgmental process: A model was introduced for an individualized normative solution which makes allowance for the subjects' information weights. The strong bias of the older subjects can at least partially be explained by these weights. The model's prediction deviates less from the graduate students' judgments than the 'normative solution'. The older subjects' information weights are nonrandom and capture the emphasis given to one or the other item of information. This may point at the active side of information processing. However, why this is done cannot be answered on the basis of our data. Whether the strong emphasis is a quick strategy which is used as no other appropriate tools are available to reduce the complexity of the situation, or whether more sophisticated arguments are hidden, remains an open question. Though we have introduced some examples for a reasonable weighting of information when referring to different interpretations of the problem and conceptions of probability, the information gained by information weights is too narrow to answer the above question, and an information weighting may also be due to other factors. Nevertheless, we want to note that weighting has evidently been identified in process studies (see the forthcoming chapters), and that younger subjects' information weights, on the other hand, do not supply a valid cue to their response behavior.

2.8. CONCLUSIONS

We are not interested in heating up the controversy about biases, fallacies, cognitive cripples, sophisticated experts, and so on. Methodological scepticism is necessary, and more knowledge is needed about how to put theory into practice, as well as about how to relate experimental findings to real life questions. We think biases and fallacies are real enough and inherent in every living system. The facts are that there is a poor level of performance in the Cab problem, its variants, and other probabilistic tasks. This is an unsatisfying situation, and in the end it is irrelevant whether we apply the bias label or not. That there is at least a communication bias or fallacy can be seen from the different interpretations and solutions given by both researchers and subjects. It is more important to find out the reasons and genesis of these decision processes, as knowledge about these reasons is a prerequisite for the development of suitable methods (e.g., curricula or decision aids) for improving behavior.

Methodological scepticism and reconstructive criticism are no substitute for hard data: Criticism is necessary. Reconstructive criticism (cf. FISCHHOFF, 1983) which points out which task representation might be possible or which concepts are suitable for handling a problem, is also indispensible. HUMPHREYS and BERKELEY, 1981 (cf. PHILLIPS, 1983) have convincingly demonstrated that a betting problem, known as the ALLAIS paradox, may vanish if risk weights are introduced. Whether the possibility of this vanishing is factual or not is an empirical question. Before this question can be answered, the internal representation of the problem and the cognitive processes of the inference should be known. For the 'base–rate fallacy', information weights are an — admittedly weak — attempt to explore subjects' problem understanding and information processing. From the conceptual side, information weights are **cognitive anticipations** similar to the aspiration levels which govern behavior in achievement situations (cf. HOPPE, 1931, p. 10; LEWIN, 1946), or bargaining (SAUERMANN & SELTEN, 1962; SCHOLZ, 1980a; TIETZ, 1983). The information weights and the model being tested in this chapter yielded a couple of findings, but they are still only crude attempts to explore subjects' problem understanding and information processing.

Guidelines for forthcoming research: Information weights are only one component in the course of problem acquisition, and only part of the variation of response distributions could be explained by these weights. If we want to grasp the individual's judgmental process, we need a lot of additional knowledge and answers to the following cascade of questions: "How is the situation perceived and framed?" (e.g., textbook vs. social judgment frame; cf. BAR—HILLEL, 1983); "How is the question understood?" (e.g., is the conditional probability converted or not); "Which meaning is given to concepts?" (e.g., does the type of probability concept matter, or are preconceptions applied by the person?); "Which tools are available?"; "How much precision is desired in the judgmental process?" (e.g., by the experimenter or by the subjects); and "What degree of effort is made by the subject to solve the problems?"

The answers to these questions are important, and the questions should be regarded as guidelines for the research in the subsequent chapters and also for future research. Of course, such research has to be accompanied by a methodological change (SCHOLZ, 1983c), as the decision process "and the forming of judgment is seen as a dynamic, social process, involving interaction between the individual and his or her environment and with other people" (cf. PHILLIPS, 1983). The paradigm of this research has now been labeled "generation paradigm" and is a challenge for decision researchers, at least if they do not disclaim the necessity of hard data.

3. A CONCEPTUALIZATION OF THE MULTITUDE OF STRATEGIES IN BASE−RATE PROBLEMS

3.1. WAYS OF UNDERSTANDING AND TREATING STOCHASTIC PROBLEMS

Although the need for an investigation of the individual's cognitive strategies has repeatedly been formulated (cf. PAYNE, BRAUNSTEIN, & CARROLL, 1978; SCHOLZ, 1981; PHILLIPS, 1983, or the critique presented in Chapter 2.3. above; positive exceptions being MONTGOMERY & SVENSON, 1983; HUBER, 1983; ALLWOOD & MONTGOMERY, 1981, 1982), very little research has been undertaken until now that may be regarded as process studies in probability judgments. The cascade of questions in the concluding section of the last chapter provides a rather broad scope of starting points for an investigation of the judgmental process. Therefore, there is a need to more closely specify the focus of this chapter. There are two issues that will be dealt with in some detail. **First,** up until the present time, hardly any knowledge has been acquired about "which comprehension of text and questions subjects have at the base−rate fallacy" (SCHOLZ, 1983a). The subjects' problem understanding should be the first step in any process study as has already been emphasized in early work on problem solving (cf. SELZ, 1913, 1922). Hence we will begin with an analysis of written reproductions of the question on the requested conditional probability and will introduce **a classification of the possible meanings** which may be attributed to the question by subjects. **Second,** the judgmental process itself will be investigated by means of an analysis of written protocols. The justifications will be analyzed in order to find out whether different **modes of thought** are elicited during probability judgments, and whether the multitude of responses given on base−rate problems may be better understood if both the subject's actual problem understanding and the mode of thought are taken into account.

Do subjects understand what we want? Empirical evidence for the difficulties subjects might have when asked for conditional probabilities is presented by BAR−HILLEL, 1984. She examined 29 college students using the Tom W. paradigm of the 'base−rate fallacy' (cf. KAHNEMAN & TVERSKY,

1973). In the original version, subjects were given a short personality description, D, of one Tom W., and were asked to rank the conditional probability, $p(H_i/D)$, of Tom W. belonging to a certain field of interest or profession, H_i. Yet in BAR−HILLEL's instructions, the subjects were asked to rank the inverse probability, $p(D/H_i)$. Surprisingly, the mean rankings of college students' answers on $p(H_i/D)$ and $p(D/H_i)$ show a correlation of .98.

What do subjects understand when asked for a conditional probability? Which questions do they answer? Which probabilities or which entities do they try to determine?

In general, every probability statement about real events contains conditional probabilities, as, to put it trivially, the population to which the statement usually refers is a subset of the universe. Further, nearly every probability textbook at secondary school level introduces the concept of conditional probability and BAYES' formula. Hence, in view of the startling problems that subjects have in dealing with conditional probability, it seems amazing that studies on the individual's understanding and subjective conceptualization of this concept are missing. Although there are substantial theoretical papers on the didactical difficulties inherent in the concept of conditional probability written by mathematicians (cf. DINGES, 1979; BOROVCNIC, 1984; RIEMER, 1981; v. HARTEN & STEINBRING, 1984; etc.), and of course, also speculations about potential (mis−)understandings contributed by psychologists (cf. BAR−HILLEL, 1980, 1983; WALLSTEN, 1983; etc.), to the best of our knowledge no direct investigations of the above question are available.

A simple procedure to find out from which understanding of the requested probability subjects actually start is to introduce various base−rate problems and to ask the subjects to rewrite the question in their own words.

In a first step, one may then classify the individuals' rewritings of questions. A framework for this classification is provided by different key concepts of the probability calculus, and different interpretations of the probability concept. Principally, subjects might try to determine a different probability than the one requested. Besides the **inverted probability** $p(D/H)$ and the diagnosticity, subjects might think that they have been asked for the **base−rates** $p(H)$.

However, the **conjunctive probability** might also be a candidate for subjects' probability judgment, as the discrimination between verbal formulations of conjunctive and conditional probability is somewhat "fuzzy". What is meant by this will be illustrated in the results section.

When trying to assess one of the above probabilities, the subjects' understanding of the questions may involve different interpretations of probability. A written reproduction of a probability question is of course only a rough indicator of the individual's understanding of probability that only provides information on a superficial language level. Hence, we will only distinguish between:
- A) formulations using the (neutral) term **probability**,
- B) **frequentistic reformulations**, such as formulations using the wordings "percentage, x out of one hundred, etc.",
- C) **subjectivistic reformulations** referring to a degree of belief or subjective confidence, and
- D) further variants, such as logical probability.

The probability concept has to be regarded as a concept which is emerging lately in the course of both ontogenetic and historical development. We cannot even suppose this concept to be universally available in the highly educated and developed "western countries" (cf. PHILLIPS, 1983). So it is also possible that subjects might not even possess a concept of probability, and that we cannot expect an appropriate probability response on base−rate problems.

Which strategies are applied? Most of the psychological explanations of subjects' performance on base−rate problems only focus on those responses which are identical with the diagnosticity. The constructs of causality (TVERSKY & KAHNEMAN, 1979), specificity or relevance (BAR−HILLEL, 1980), or vividness (NISBETT & ROSS, 1980) for example, conceptualize the subject's complete neglect of base−rates with reference to certain content or context parameters (cf. Chapter 2.2.).

The 'normative solution' $p(H|D)$ of the Hit Parade problem generally lies somewhere between the base−rate $p(H)$ and the diagnosticity $p(D|H)$ (we are only considering the so−called **normal version** of variants **of the Cab problem** here in which $p(H) < p(H|D) < p(D|H)$). Responses within this interval are usually explained either by the anchoring and adjustment hypothesis (cf. NISBETT & ROSS, 1980, p. 41), or information weights (cf. Chapter 2. or SCHOLZ & BENTRUP, 1984). The anchoring and adjustment hypothesis states that people stick to the salient features of the task and show (insufficient) adjustment in the direction of the 'normative solution'. However, both the anchoring and adjustment hypothesis, and models based on information weights, can only explain a part of the variation of responses made by the subjects. It would appear that probability response behavior shows a wider range of strategies or coping behavior than has previously been assumed.

3.2. THE ANALYTIC AND THE INTUITIVE MODE IN STOCHASTIC THINKING

In a sequence of preliminary experiments, various techniques were applied in an attempt to conceptualize the multitude of strategies in base—rate problems. Subjects' processing was traced by listening to series of thinking aloud protocols. Postexperimental interviews were conducted on the subjects' proceeding and interpretation of the problems. Furthermore, various taxonomies of strategies were tentatively developed when referring to the operations which were inherent in the calculations explicated in the written protocols (from Experiment B, reported below). After consulting various methods and facing more than two hundred protocols and interviews, the author hypothesized that **different modes of thought** guide the judgmental process. We propose that if we want to understand the cognitive process of probability estimation, it is essential to distinguish between intuitive and analytic thinking. The complementary concepts of intuitive and analytic thinking have not only been applied in the psychology of thinking. A variety of conceptions of intuition have been developed in a range of contexts. The mathematician FISCHBEIN, 1983, for instance, separates **primary intuitions** which he regards to be completely unrelated to instruction and education, from **secondary** intuitions which "convey the products of social experience in the form, mostly, of scientific truths." (FISCHBEIN, 1975, p. 9). In another paper (FISCHBEIN, 1982), he distinguishes between anticipatory and affirmatory intuitions. WESTCOTT, 1968, pointed out that the different foundations for the definition of intuition in various sciences depend on the different problems to be dealt with. Philosophers have used the concept of intuition to grasp the nature of knowledge, psychologists and educators for understanding personality in clinical practice, or for conceptualizing learning processes.

We want to clarify that, unlike philosophers, we are not concerned with intuitionism as a form of knowledge. Our interest is in **intuition as a mode of, and also as the "product of accepted psycho—physiological processes of thought and behavior that occur under particular conditions of personality, environment, and experience"** (BASTICK, 1982, p. XXIII). Hence, from our cognitive decision theoretical point of view, we regard intuition as a natural and common mode of thought (cf. OTTE & JAHNKE, 1982) which is, of course, an object of development and education like every other form and product of thinking. We will also distinguish **intuition** from **insight**. Insight will be regarded as a task—related 'cognitive reorganizing' ability that permits an understanding of the causes of a problem or its solution. As NEISSER, 1963, emphasized, the components of insight must already have been established before a task or problem can be treated.

Intuition has often been given a very negative definition. BUNGE, 1962, p. 113, denotes intuition as "rough" and "dangerous", and less psychological research has been devoted to this concept than one might expect. Although, or maybe even because, "intuitive and analytical thinking are common, everyday facts of life, ... psychologists neglected ... the study of the relations between intuition and reason: Everyone 'knew' and consequently there was no need for empirical studies" (EARLE, 1972, p. 71). The unwanted consequences of this were that "... we know virtually nothing about this relationship". Today, this situation has scarcely changed. We will introduce a definition of modes of thought by lists of features which may be regarded as descriptions or definitions of the prototypical intuitive and analytic modes of thought. The presented lists are a kind of intermediate—product in the spiralling process of the conceptualization of the strategies which was accompanied by a permanent oscillation between data and theoretical studies. Both the composition of the list and the brief description of the single features and attributes have been developed in this process. Consequently, some references may not have been credited in each appropriate place. Therefore, we wish to note that the subsequent conceptualization has been constructed with reference to the BRUNSWIKian research by HAMMOND and coworkers (cf. HAMMOND, McCLELLAND, & MUMPOWER, 1980; HAMMOND, HAMM, GRASSIA, & PEARSON, 1983), and BASTICK's, 1982, detailed volume on intuition. Further references which have influenced the proposed definition are: BERNE, 1949; BRAINE, 1978; BROMME & HÖMBERG, 1977; BRUNER & CLINCHY, 1966; CLINCHY, 1975; NEISSER, 1963; ORNSTEIN, 1976; POINCARE, 1929/1969; SKEMP, 1971; SINZ, 1978; WACHSMUTH, 1981; WESTCOTT, 1968; and WESTCOTT & RANZONI, 1963. KUHL's, 1983a, b; and KAHNEMAN and TVERSKY's definitions of intuitive thinking should also be mentioned although they are less encompassing and also not completely contained in the subsequent proposal. For example, KAHNEMAN and TVERSKY, 1982b, p. 494, simply give the definition, "..., a judgment is called intuitive if it is reached by an informal and unstructured mode of reasoning, without the use of analytic methods or deliberate calculation."

Table 3.1.: List of attributes and features of intuitive and analytic thought

	Intuitive thought	Analytic thought
A	preconscious — information acquisition — processing of information	conscious — information acquisition and selection — processing of information
B	understanding by feeling and instinct of empathy	pure intellect or logical reasoning, independent of temporary moods and physiology
C	sudden, synthetical, parallel processing of a global field of knowledge	sequential, linear, step–by–step ordered cognitive activity
D	treating the problem structure as a whole, "Gestalt erkennend"	separating details of information
E	dependent on personal experience	independent of personal experience
F	pictorial metaphors	conceptual or numerical patterns
G	low cognitive control	high cognitive control
H	emotional involvement without anxiety	cold, emotion–free activity
I	feeling of certainty toward the product of thinking	uncertainty toward the product of thinking

CHAPTER 3.2.

Features and attributes of analytical and intuitive thinking

A. Preconscious vs. conscious. We will distinguish between the conscious and the preconscious acquisition and processing of information (cf. HOGARTH, 1980). Acquisition of information is preconscious if the individual is not aware of the process of information input and hence does not memorize the source and genesis of information. If a conclusion is made on the basis of less information than required to reach this conclusion (cf. CLINCHY, 1975), or no adequate account is given for the decision (cf. THORSLAND, 1971), or if arguments are not completely explicit or well−grounded, we speak of preconscious information processing. Oppositely, conscious information processing is characterized by conclusive and reflective arguments, and in conscious information acquisition the individual is aware of the information input.

B. Understanding by feeling vs. pure intellect. Reason and pure intellect are impersonal, logical, independent of physiology (e.g., emotional arousal, cf. BASTICK, 1982), highly reliable in repetition, and predictable (cf. HAMMOND, 1984). This can be contrasted with understanding by feeling or empathic instinct. Such intuitive understanding is often indicated by personal speech and is dependent on temporary moods and states of physiology.

C. Sudden, synthetical, parallel processing of a global field of knowledge vs. sequential, linear, step−by−step ordered cognitive activity. Analytic thinking often features the step−by−step connection of arguments which are taken from one level or frame. Sometimes this can also have negative consequences, if the status and significance of the step−by−step operations are not reflected. Linear deductive sequential chains, mechanical if−then connections, or quasi−blind, empty, and senseless applications of algorithms are features of such analytical thought (cf. WERTHEIMER, 1959, p. 10).

D. Treating the problem structure as a whole ("Gestalt erkennend") vs. separating details. Intuitive activity is the treating of the problem structure as a whole, avoiding decomposition or separation into isolated pieces of information. The additional label, "Gestalt erkennend" is chosen, as not only the wholeness of the activity is taken as a criterion, but also a certain degree of penetration into the structure of the problem or situation must be involved. On the other hand, the separation of elements, the recognition and analysis of details, and a complete listing of information are attributes of analytic thinking.

E. Dependent vs. independent of personal experience. Intuition has to be developed and relies on past personal experience that has often been gained through a direct situational involvement. The addition of information and speculative reevaluations based on personal experience are on the intuitive side, whereas the analytic side is impersonal and restricted to informations which are directly given by the situation or task.

F. **Pictorial metaphors vs. conceptual or numerical patterns.** In the intuitive mode of thought, pictures or images of similar situations are often recalled, and the inferential process is guided by the experience gathered in these situations. Visualization of a problem (for instance in geometrical problem solving) is also a feature of the intuitive mode of thought, whereas the use of theoretical concepts (such as geometrical theorems or stochastic independence, negation of negation in base−rate problems, etc.) or numerical and quantitative patterns are attributes of the analytic mode of thought.

G. **Low vs. high cognitive control.** Analytic thinking is a controlled activity that checks the steps of the inferential process and results. The control arguments, however, are usually taken from the same level as the arguments themselves. For instance, a division of two numbers is controlled by a remultiplication. Intuitive thinking lacks this step−by−step checking. The thinker arrives at an answer with little if any control of the process of thinking (cf. BRUNER, 1960).

H. **Emotional involvement vs. cold, emotion−free activity.** Intuitive thinking is mostly coupled with an emotional involvement (this is the primary message of BASTICK's, 1983, research work on intuition). Yet this involvement is free of anxiety and fear and thus does not impede the cognitive activity. On the other hand, analytic thinking is accompanied by cold and self−protective emotion, a state that is often labeled as emotion−free activity. KUHL, 1983b, p. 238, for instance, argues that emotions signalling danger or discomfort elicit the analytic mode.

I. **Feeling of certainty vs. uncertainty toward the product of thinking.** BASTICK, 1982, pp. 21, states that analytic thinking does not result in a feeling of certainty, as it is devoid of feeling. Intuitions, however, elicit a feeling of certainty, even if they are wrong or if they lead astray. HAMMOND et al., 1983, p. 9, differentiate between confidence in the answer, but not the method, as in intuition, and confidence in the method, but not the answer, as in analysis. HAMMOND's point of view seems to be not only more differentiated, but also more adequate, when one recalls the certainty of mathematics students toward their performance when they have finished running through a complicated calculation or proof ending up with an unlikely result.

Various remarks have to be made on the list of features and the modes of thought, their structure, their genesis, and their intended use.

Of course, stochastic thinking does not only occur in the ideal forms of intuitive and analytic thinking. Thus we do not believe in a discrete dichotomy between the modi. We do not completely agree with HAMMOND et al., 1983, p. 9, who obviously regard the modes to be a unidimensional construct: "cognitive activity is not a dichotomy of intuition and analysis, but rather a continuum marked by intuition at one pole and analysis at the other". According to the list of features the modi are **multidimensional constructs**, so we regard the introduced modi as contrasting forms of cognitive activity. We speak of a prototypical intuitive/analytic mode of thought if all features or attributes from the analytic/intuitive side are attributed as in a protocol on thinking. We propose that there are intermediate forms of thought, non-classifiable cases (for instance due to a consecutive sequence of rapid changes between the modes), and also other forms of thought that may be named just as the intuitive and the analytic mode have been named.

Naturally, sometimes it might be difficult to discriminate between the modes according to the dimensions. If, for instance, the object of thinking is one's own personal experience, or emotions, or geometrical objects, one might be tempted to discriminate between the modes according to the object of thinking, and one might thus be tempted to assign an intuitive feature in the case of an analytic proceeding. Nevertheless, we suggest that the individual strategies in base—rate problems may be roughly judged to be on the analytic or on the intuitive side.

Clearly the dimensionality of the modes of thought is not the number of features. As has been mentioned above, the list has to be viewed as an interproduct of theoretical considerations, documented empirical findings, and the experimental evidence which has been provided by a series of pilot studies. Thus, on the one hand, the interdependent relationships of the features and the dimensionality of the list have to be clarified by empirical studies. This may lead to a modification of the list and a reduction of its dimensionality. On the other hand, the list should be supplemented and refined by further studies. Nevertheless, we believe the features to be independent in the sense that they may be varied pairwise so that pairwise all combinations of intuitiveness and analyticity may result (though this statement does not reveal anything about the dimensionality of the scale). Naturally the concrete structure of the list depends on its genesis and intended applications. The list may be considered as an **operationally—oriented definition of modes of thought** that may be applied to verbal or written protocols of the process of thinking. Some features that might be identified when using video or eye movement recordings (like hypnotic reverie) thus will not be included.

Algebraic vs. nonalgebraic and analytic vs. intuitive strategies. KAHNEMAN and TVERSKY, 1979a, 1982b, or EARLE, 1971, 1972, propose that in probability judgment problems, analytic thinking implies algebraic calculations, and intuitive thinking a nonalgebraic processing. However, in our description, only Feature F entails numerical patterns as one possible analytic attribute. On the other hand, we know that in mathematics algebraic operations are sometimes applied preconsciously and intuitively. Hence, we also expect algebraic—intuitive and nonalgebraic—analytic strategies to exist. However, we also anticipate a correlation between the modes of thought and the application of algebraic operations, as in the context of probability judgments by word problems, intuitive strategies are supposed to be mostly free of algebraic operations, whereas analytic strategies are often accompanied by algebraic operations.

The **aim** of the following **Experiment B** is to investigate whether the different modes of thought are present in probability judgments in base—rate problems. We will examine whether the definition of an intuitive vs. analytic mode of thought provides a reliable rating procedure. The modes of thought are not only supposed to be of theoretical interest, but also to result in different response distributions. This is why we conducted an exploratory data analysis on the distributions that result from the two modes of thought.

3.3. EXPERIMENT B

3.3.1. SUBJECTS

The subjects were 88 student volunteers from a German Gesamtschule (comprehensive high school) and 30 postgraduate students of Bielefeld University who had received their Vordiplom (a pre—diploma, received after two years of university studies). The high school students were of different age levels. There were 17 seventh grade students, 31 ninth or tenth grade students, and 40 twelfth grade students.

3.3.2. PROCEDURE AND EXPERIMENTAL TASKS

The questionnaire included the Hit Parade, the TV, the Motor problem, and one non−Bayesian problem with different orders of presentation. The base−rate problems were the same as those reported in Chapter 2.1. On the last pages of the questionnaire, the first problem was repeated, and the subjects were asked to reformulate the question (i.e., to write down the question in their own words), to give another probability judgment in percentages, and to write down an extensive justification of their judgment process, that means, to provide a protocol on their thought process. There was a time limit of one hour for the questionnaire. It should be noted that while the subjects were reformulating and answering the questions, and justifying their answers, the text was at all times visible to them.

3.3.3. METHODOLOGICAL REMARKS ON THE USE OF WRITTEN PROTOCOLS

The data basis for the analysis of the cognitive strategies is the **written justifications**. This use of a written justification to deduce conclusions about the cognitive process has been criticized for several reasons. We will briefly describe the most essential ones.

Limitations in the ability to report thought processes: It is well−known that people are not able to report all their internal states when performing a process: Previously heeded information may already have disappeared from short−term memory (cf. ERICSSON & SIMON, 1980, p. 226), the instruction to articulate or to justify thinking may distort the processes (cf. POSNER, 1982), and subjects may report how they think they ought to think, but not how they actually do think (cf. NISBETT & WILSON, 1977).

Of course, verbal or written protocols are not the cognitive processes themselves but have to be interpreted. Even proponents of protocol analysis like ERICSSON and SIMON, 1980, 1984; PAYNE, BRAUNSTEIN and CAROLL, 1978; or HAYES, 1982, have pointed out that not all mental processes can be verbalized. For example, moral judgments based on non-conscious prejudices, or motor processes that do not involve short−term memory. Presumably, it is seldom that the written justifications (and also the verbal protocols!) run concurrently and simultaneously to the mental processes being reported. For instance, thinking, speaking, and writing are governed by different speed limitations. Hence, the user of protocol analysis cannot exclude the possibility that important steps of the judgmental process may eventually even be systematically omitted in the written justification, for example, "relevant information may be lost from short–term memory" (HAYES, 1982, p. 67), or maybe some subjects give what they regard to be an appropriate answer post hoc. However, despite these criticisms, protocol tracing is a research method which "can provide insights into underlying processes" of a judgment, choice, or decision and "... can be used to test hypotheses or alternative theories" (LEWIN, 1982, p. 315), and these are the objectives of this study. Positive results on the use of retrospective reports in the investigation of causes of decisions are also given by WRIGHT and RIP, 1981.

Simultaneous reporting may affect the mode of thought: Most of NISBETT and WILSON's, 1977, much quoted criticism of the use of protocol analysis was based on an analysis of retrospective reports. This may be why many proponents of verbal protocol analysis consider concurrent reporting to be the only valid method. Others, like LICHTENSTEIN, 1984, p. 84, have pointed out that both techniques "have value and both have flaws". We agree with LICHTENSTEIN, and believe that no single technique is generally superior in documenting the crucial phases of thought processes. Sometimes the thought process may be distorted by concurrent reporting and the subjects may prefer to report retrospectively. On other occasions, the simultaneous sketching of loosely associated ideas and impressions may provide better insights than necessarily incomplete post hoc reconstructions. But we must not forget that there are also some objections to the exclusive use of the technique of simultaneous thinking aloud. The most important argument has been formulated by KUHL, 1983b, p. 237, who proposes that the request to think aloud induces a shift toward the analytic mode of thought. We would argue that if this proposition is true it would be even more applicable to written protocols. (In a further study, cf. SCHOLZ & KÖNTOPP, in prep., we will systematically control the effect of simultaneous vs. retrospective protocoling.) Hence, we presume that on the basis of these arguments, in a first step, the best solution for our research purpose would be to inform the subjects about the aim of the experiment, which is to trace and to understand the individual's particular thinking process, and then allow them to decide which type (concurrent vs. retrospective) of recording they judge to be more appropriate to convey their thought processes. We also do not consider that our subjects' reports are markedly biased by a social desirability shift, as we share the same experience as LICHTENSTEIN who reported that subjects are "willing to be utterly honest, even when that meant that they looked quite foolish" (LICHTENSTEIN, 1982, p. 84).

3.3.4. MEASURES AND RATING PROCEDURE

The **reformulations of the question** and the **written justifications** were the object of a content analysis. The categories of the analysis were developed when facing the data reported. Hence any inferential statistics have to be interpreted with restrictions.

Any classification of longer verbal or written texts into semantic categories is fuzzy. In order to provide at least a minimum of objectivity, the classifications into the categories that were discussed in Chapter 3.1. (cf. also Table 3.2. below) were independently performed by the author and one other rater. Diverging classifications of the **reformulations of the questions** were discussed until a consensus was reached.

The protocols of the justifications were classified according to the nine features of intuitive and analytic thinking. Each dimension received an intuitive, analytic, or nonclassifiable/intermediate rating. The coder instructions contained the above descriptions of the features and attributes of the two modes of thought, some prototypical protocols as examples, and a prescribed order of coding. In order to avoid both fatigue from consecutive coding of one and the same feature more than one hundred times, and unjustified high correlations between the single features and attributes due to a sequential coding of all features on one and the same protocol, the coders were asked to rate four protocols in turn, feature by feature. We wish to note here also that a strategy is labeled intuitive according to a rater's protocol coding, if from the nine features, at least two more features are rated intuitive than analytic. The same procedure in reverse was used to rate the analytic strategies. The ratings on the modes of thought were performed independently by the author (Rater 1) and three other raters (Rater 2, Rater 3, and Rater 4).

3.3.5. RESULTS

Reproduction of questions. Table 3.2. presents the absolute frequencies for the classifications of subjects' reformulations of the questions. Approximately half of the subjects reproduced texts which are rated as correct reproductions. Most

PART I

of these reproductions are literal, only slightly modified copies of the original questions using the term 'probability'. Two issues should be regarded when considering this result. First, although the subjects were asked to reformulate

Table 3.2.: Categories and frequencies of reproduction of questions in the different stories

Categories		Motor	Hit	TV	All problems
I	Correct reproduction	22	18	25	65
IA:	Almost literal	20	13	23	56
IB:	Frequentistic	-	1	1	2
IC:	Subjectivistic	1	-	-	1
ID:	Further variants	1	4	1	6
II	Inverting cond. prob.	4	11	2	17
IIA:	Using 'probability'	2	1	2	5
IIB:	Frequentistic	1	5	-	6
IIC:	Subjectivistic	-	-	1	1
IID:	Further variants	1	4	-	5
III	Base-Rates	-	2	3	5
IIIA:	Using probability	-	1	2	3
IIIB:	Frequentistic	-	-	1	1
IIID:	Further variants	-	1	-	1
IV	Conjunctive prob.	-	3	1	4
IVA:	Using probability	-	3	1	4
V	Nonprob. interpret.	2	-	3	5
VI	No paraphrasing of questions	2	1	-	3
VII	Not definitely classifiable statements	10	2	3	15
		40	40	38	118

the questions in their own words, a considerable number of them obviously just copied the visible question, and second, some of the subjects assigned to Category I (correct reproduction) did not necessarily have to reformulate a conditional probability, as in the Motor problem, the condition (by which the probability space has been restricted) is part of the description of the problem.

Few reformulations are assigned to the subcategories B) frequentistic, and C) subjectivistic reformulations (c.f. Table 3.2.). There are six formulations in category ID. These formulations are rated as correct, as they reproduce the question on a conditional event, but use an idiosyncratic, nonstandard, or neutral formulation of the probability concept (e.g., how big is the chance), which could not be definitively classified. We present an example of the latter, supplied by a 24−year−old postgraduate student. All protocols introduced in the subsequent text will be presented in the original German wording alongside an English translation due to the difficulties of direct literal translation.

"The question is, how big is the studio guest's influence on the listener who votes "hit" for a song recommended by the studio guest."

Es wird danach gefragt, wie groß der Einfluß des Studiogastes auf den Zuhörer ist, der einen vom Studiogast vorgeschlagenen Schlager als Hit tippt.

About 14% of all subjects inverted the conditional probability and produced formulations of questions on the diagnosticity. Formulations using the term probability, frequentistic and subjectivistic formulations, and further variants can likewise be found. We give an example from a 19−year−old male subject:

"What is sought is a percentage. That is, if the case given in the question (i.e., random title, studio guest, and listener) occurs one hundred times; how often would the studio guest be able to predict the listeners' taste."

Gefragt ist nach einem Prozentsatz. Das heißt, wenn der in der Frage gestellte Fall (also zufälliger Titel, Studiogast und Hörer) ein hundert mal eintritt: Wie oft würde der Studiogast den Hörergeschmack voraussagen können.

It was most difficult to discriminate between formulations of the requested conditional probability and the conjunctive probability $p(H \cap D)$. Verbal formulations of both probabilities may be quasi−continuously modified. We will demonstrate this through some of our data. The following formulation is given by a 21−year−old, male, 12th grade student. He wrote:

"The question is to determine the probability for the same judgment of a title, that means the same judgment by the studio guest and by the listeners and that it will be a hit."

Die Frage ist die Wahrscheinlichkeit der gleichen Beurteilung eines Titels, von Hörern und Studiogast, und zwar als Hit, herauszufinden.

The classification of the next text, produced by a 19−year−old male 12th grader, is less unique and could not be classified by the raters.
> "How big is the probability that the studio guest judges a title to become a hit and it also will become one?"
> Wie groß ist die Wahrscheinlichkeit, daß der Studiogast einen Titel als Hit beurteilt und er auch einer wird.

At least if the word **then** is inserted in the second phrase, the border to adequate formulations of the diagnostic probability may be crossed. Such a reformulation is given by a 16−year−old female 10th grader:
> "How big is the probability that a chosen song will be named as a hit by the studio guest, and then also by the listeners."
> Wie groß die Wahrscheinlichkeit ist, daß ein gesuchter Titel von dem Studiogast als Hit benannt wird und dann noch von den Hörern.

Five reformulations of questions did not refer to a probability statement (Category V), and 3 subjects did not paraphrase a question, but gave some other statement. Lastly, 15 statements could not be definitively classified.

Before turning to the analysis of the strategies, we will note that 9 out of the 17 subjects from Category II actually responded with the diagnosticity, and just 1 out of the 5 subjects who reformulated a question for the base−rates produced a base−rate response. As has already been mentioned in the procedure section, we are here reporting on a second response on the same base−rate problem within a questionnaire. The response distributions on the various problems of this second response did not differ significantly from those obtained for the first answer, and are thus extremely similar to those reported in Chapter 2 (cf. Figure 2.5).

Interrater reliability on the classification of modes of thought. The method of classifying strategies by protocol rating may be considered to be a weak method. Furthermore, the necessity for a detailed scale analysis has been quoted above. This is why we will present the interrater reliability analysis in an extensive manner as it is above all crucial for an application of the list of features. We will first present the distributions of ratings by all raters for each single feature and for the classification of strategies. As may be seen in Table 3.2., the coders show different tendencies toward analytic or intuitive scores. For instance, Rater 3 is more prone to provide ratings resulting in analytically classified strategies than the other raters. The ratios of analytic to intuitive strategies are 4.25 for Rater 3, and 1.9, 2.1, and 1.1 for Raters 1, 2, and 4. The different characteristic of Rater 4 has some implications on the evaluation of descriptive measures, e.g., the percentage of matching ratings (including the nonclassifiable/intermediate category), and the percentage of matchings of classified ratings, or the Z−value. The Z−value is an appropriate, and well approved descriptive measure for judging the interrater reliability (cf. LISCH & KRIZ, 1978; RITSERT, 1972). It is defined as follows:

$$Z_{i,j} = 2 \cdot M_{i,j} / (C_i + C_j)$$

where $M_{i,j}$ is the total number of matchings between rater i and j, and C_k the total number of analytic−intuitive ratings by rater k, k=i,j. According to an accepted rule of thumb, a Z−value above .8 may be considered to indicate a high interrater reliability. Theoretically, the Z−value ranges between 0 and 1 with an expected value of .33 if independence and a uniform marginal frequency distribution is assumed.

Table 3.3.: Frequencies of analytic and intuitive scores for the Features A to I, and the strategies according to the raters' judgments (N = 118)

Features	A	B	C	D	E	F	G	H	I	$\bar{x}$	Strat.
Rater					Analytic scores						
1	62	48	50	39	78	50	46	49	42	51.6	69
2	60	65	45	57	77	28	31	64	31	50.9	75
3	57	54	51	27	103	83	57	96	15	60.7	85
4	50	51	64	25	86	52	28	57	9	41.3	56
					Intuitive scores						
1	43	52	37	34	20	16	44	14	24	31.6	37
2	38	36	24	22	18	12	27	14	12	22.6	36
3	41	21	24	23	9	3	27	5	71	24.9	20
4	57	58	32	45	17	13	50	11	30	34.8	49

We will demonstrate the implications that the differential proneness of the raters to produce an analytic or intuitive rating have on the reliability measure exemplarily for the Z−value on the strategy classification. Let us assume, for this sake, that all raters perfectly match in the rank−ordering of all 118 subjects in respect to the analyticity−intuitiveness scale. However, the bounds of classification differ. This means, for example, that one rater may judge only 17% of the subjects to process intuitively and 72% analytically, as Rater 2 did, whereas another rater may judge 42% to process intuitively and only 47% analytically, as Rater 4 did. This results in a Z−value of .84. Because this Z−value is maximal for the exemplarily chosen frequencies of ratings, it will be denoted as Z_{max} and is an upper bound of the Z−value for the above frequencies of categories. In general, Z_{max} is defined as follows:

$$Z_{max}(i,j) = \frac{\min(C_i^a, C_j^a) + \min(C_i^i, C_j^i)}{C_i + C_j}$$

where C_k^a denotes the number of analytic, and C_k^i the number of intuitive ratings for $k = i,j$, and $C_k = C_k^a + C_k^i$.

We will thus also list the Z_{max} value at least for the strategy ratings. In the case of the matching probability, we will report the Chi^2 value which provides information about the contingency of the ratings when weighting the marginal distributions. The critical Chi^2 values for the .01 and .001 levels of significance and three degrees of freedom are 13.3 and 18.5.

Table 3.4.: Percentage of matching scores in strategy rating; a) including nonclassifiable/intermediate ratings, and b) only considering classified strategies

a) Rater	1	2	3	4	b) Rater	1	2	3	4
1	-	.81	.73	.65	1	-	.94	.84	.77
2	.81	-	.74	.66	2	.94	-	.90	.80
3	.73	.74	-	.61	3	.84	.90	-	.76
4	.65	.66	.61	-	4	.77	.80	.76	-

CHAPTER 3.3.5.

Table 3.4. presents the percentage of matching scores in strategy ratings with and without the classified strategies, whereas Table 3.5. presents the Chi^2 and the Z−values for the single features and the strategies. Additionally, the Z_{max} values are inserted in the last column of Table 3.5. It is well known that Chi^2 analysis should not be applied if more than 20% of the cells show an expected cell frequency smaller than five (cf. e.g., SIEGEL, 1976), as unjustifiably large Chi^2 values may result from these cells. Hence some of the values are placed in brackets and should be interpreted with reservations.

Table 3.5.: Chi^2 values for pairwise matchings of raters including non-classified ratings and Z−values (a Chi^2 value in brackets indicates that more than two out of the nine cells show an expected frequency of less than five; the Z_{max} values are inserted in the last column)

Features	A	B	C	D	E	F	G	H	I	Strat.
Pairs of raters				Chi^2 values including nonclassified ratings						
1 x 2	44.7	75.3	56.0	30.7	78.0	13.4	69.6	69.8	13.6	(74.8)
1 x 3	30.2	31.5	23.9	24.8	(30.0)	(60.5)	44.0	(21.8)	2.3	50.0
1 x 4	6.7	28.9	32.3	5.9	43.8	37.6	15.2	45.4	.9	33.4
2 x 3	23.1	43.0	28.7	24.4	(58.5)	(33.8)	48.6	(27.4)	4.6	(59.4)
2 x 4	17.9	30.4	20.8	8.8	49.2	27.8	17.9	23.4	9.4	(39.7)
3 x 4	23.8	29.9	32.8	10.3	(75.6)	(43.6)	24.0	(23.1)	4.6	(34.7)
				Z-values						Z_{max}
1 x 2	.74	.80	.69	.57	.85	.40	.69	.75	.42	.87	.97
1 x 3	.68	.57	.54	.44	.78	.61	.64	.61	.28	.78	.84
1 x 4	.54	.66	.66	.40	.78	.63	.52	.64	.19	.70	.78
2 x 3	.61	.67	.47	.44	.81	.41	.51	.71	.11	.81	.88
2 x 4	.60	.68	.57	.36	.78	.48	.41	.63	.24	.72	.84
3 x 4	.64	.60	.63	.35	.87	.65	.49	.66	.35	.68	.72
All pairs	.64	.66	.59	.43	.81	.53	.54	.67	.27	.76	

We will now discuss the **interrater reliability of the strategy classifications**. The average Z−value across all ratios is .76 which is slightly below the critical .8 value. All Chi^2 values for strategy matching are well above 18.5. As may already be perceived by the Z−value, the Chi^2 values, and the other measures, Rater 4 shows a rather low interrater reliability compared with the other raters. If Rater 4 is excluded, the average Z−value is .82, and above the critical value.

In order to provide some insight into the Z−measure and the other measures, we will exemplarily present the 3 x 3 table of Rater 3 and 4's matchings (cf. Table 3.6.) which, with a value of only .68, is the worst interrater reliability to be observed on the data reported. None of the 20 intuitive classified strategies of Rater 3 were rated as analytic by Rater 4, and none of the 56 analytic strategies of Rater 3 were rated as intuitive by Rater 4. At a first glance the total of 22 mismatchings (Rater 4 — intuitive x Rater 3 — analytic) seems to be a very high number, but one has to take into consideration the differential frequencies of category ratings. For instance, there is theoretically a minimum of 16 mismatchings in the intuitive x analytic cell given Rater 3 and Rater 4's marginal distributions. Thus the relatively low Z—value is predominately caused by the differential frequency of category rating.

Table 3.6.: Strategy classification matchings for Rater 3 x Rater 4

	Rater 3			
	Intuitive	Intermediate/ nonclassified	Analytic	Sum
Rater 4				
Intuitive	19	8	22	49
Intermediate/ nonclassified	1	1	11	13
Analytic	0	4	52	56
Sum	20	13	85	118

Another view of the similarity of the raters' codings may be gained by the application of cluster analysis. Within this text we will restrict ourselves to hierarchic—agglomerative procedures. The clustering within the latter, however, depends on the criterion for group similarities which usually cannot be derived from content—related considerations (cf. SCHNEIDER & SCHEIBLER, 1983; SCHOLZ, SEYDEL, RECHBAUER, & WENTZ, 1983, p. 284). Thus, we have to rely on the recommended exploratory procedure of varying the criterion and looking at the robustness in respect to this variation. Using the EUCLIDian distance measure, the Furthest, and the Nearest Neighbour, the Group Average, Wards, McQuitty's, and the Lace Williams procedures were calculated for both raters using WISHART's, 1975, CLUSTAN program.

CHAPTER 3.3.5.

If the raters' codings of the 118 subjects are taken as item vectors with, for example, an intuitive rating coded as 0, an intermediate/nonclassifiable as 1, and an analytic rating as 2, then the similarity of the raters' coding is reflected by the level of clustering in the dendrogram. As may be concluded from Figure 3.1., the relatively bad reliability of Rater 4 is also reflected by an outlier characteristic in the dendrograms, whereas Raters 1 and 2 show the greatest similarity. For the sake of completeness, we want to note that all agglomerative methods produce clusters that are the same as those produced by the Group Average method presented above.

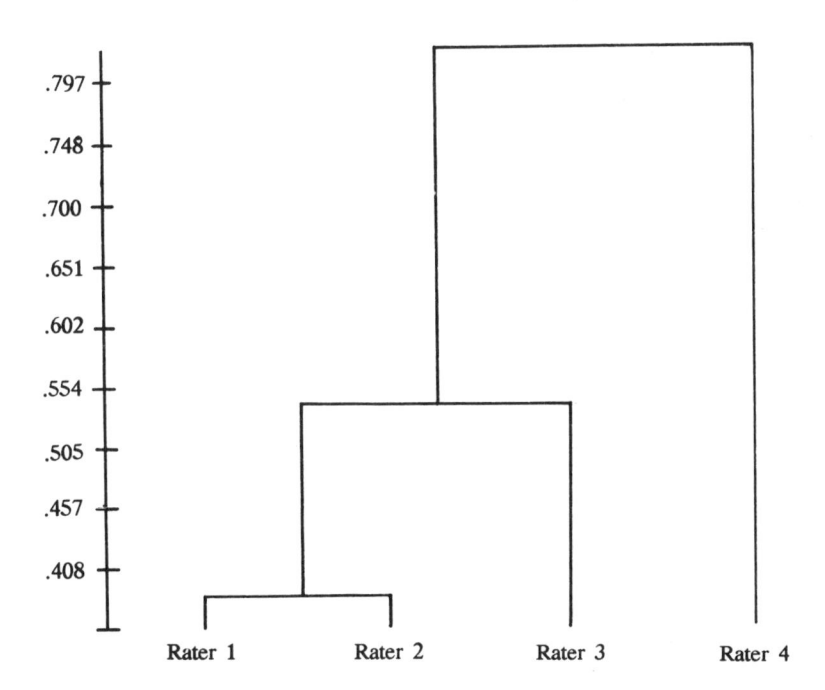

Figure 3.1.: Dendrogram of rater clustering using WARD's method (see text; y−axis: minimal values of group variance)

We will now turn to the individual features. The average Z−value for all but one feature is below the average Z−value of strategy matching. Only Feature E (dependent vs. independent of personal experience) is above .8. However, this is at least partially due to the extremely low number of intuitive ratings. Two features bring about (extremely) low interrater reliabilities. These are Features D (treating the problem structure as a whole vs. separating details) and I (confidence). The matchings of Feature D are only slightly above chance level for four pairs of raters, whereas the average Z−value of Feature I is even below chance level for all but one pair. In order to investigate the impact of these two features' sloppy reliability on the Z−value of the strategy classification, we excluded them from the strategy classification procedure. However, unexpectedly, the Z−value and the other measures of reliability are on average slightly lower if either Feature I, Feature D, or both are excluded.

The structure of the intuitiveness−analyticity scale: Some interesting information about the structure of the intuitiveness−analyticity scale can be gained by looking at the multivariate dependencies between the single features. The cluster analysis method is again a suitable procedure for describing similarities between the features' ratings in our sample. If the coders' ratings on the single features across the 118 subjects are taken as vectors, the similarities between the dimensions of the intuitiveness−analyticity scale can be reflected by the level of clustering of these vectors. We will follow the above guidelines on the cluster analysis and briefly describe the stable clusters for each of the four raters and the Majority judgment of all raters.

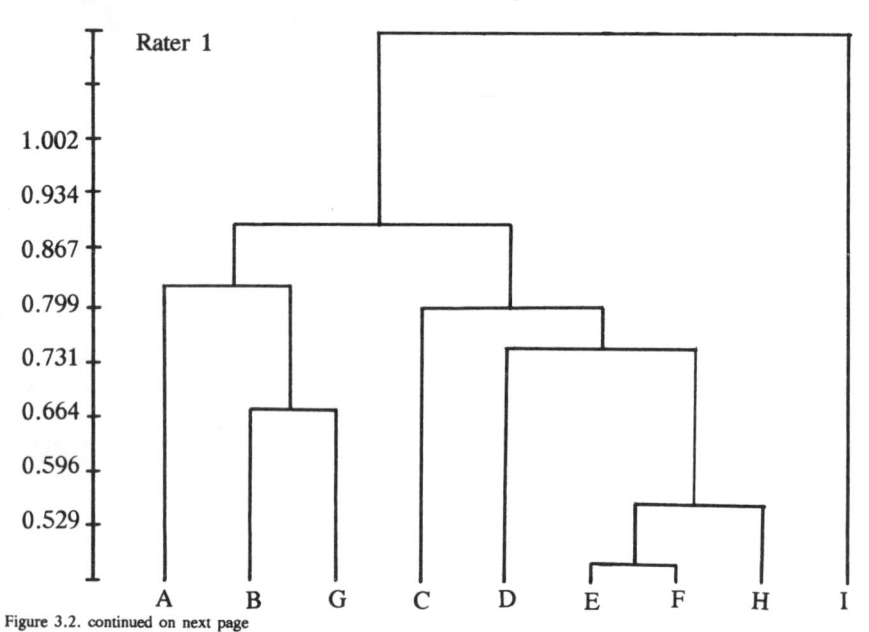

Figure 3.2. continued on next page

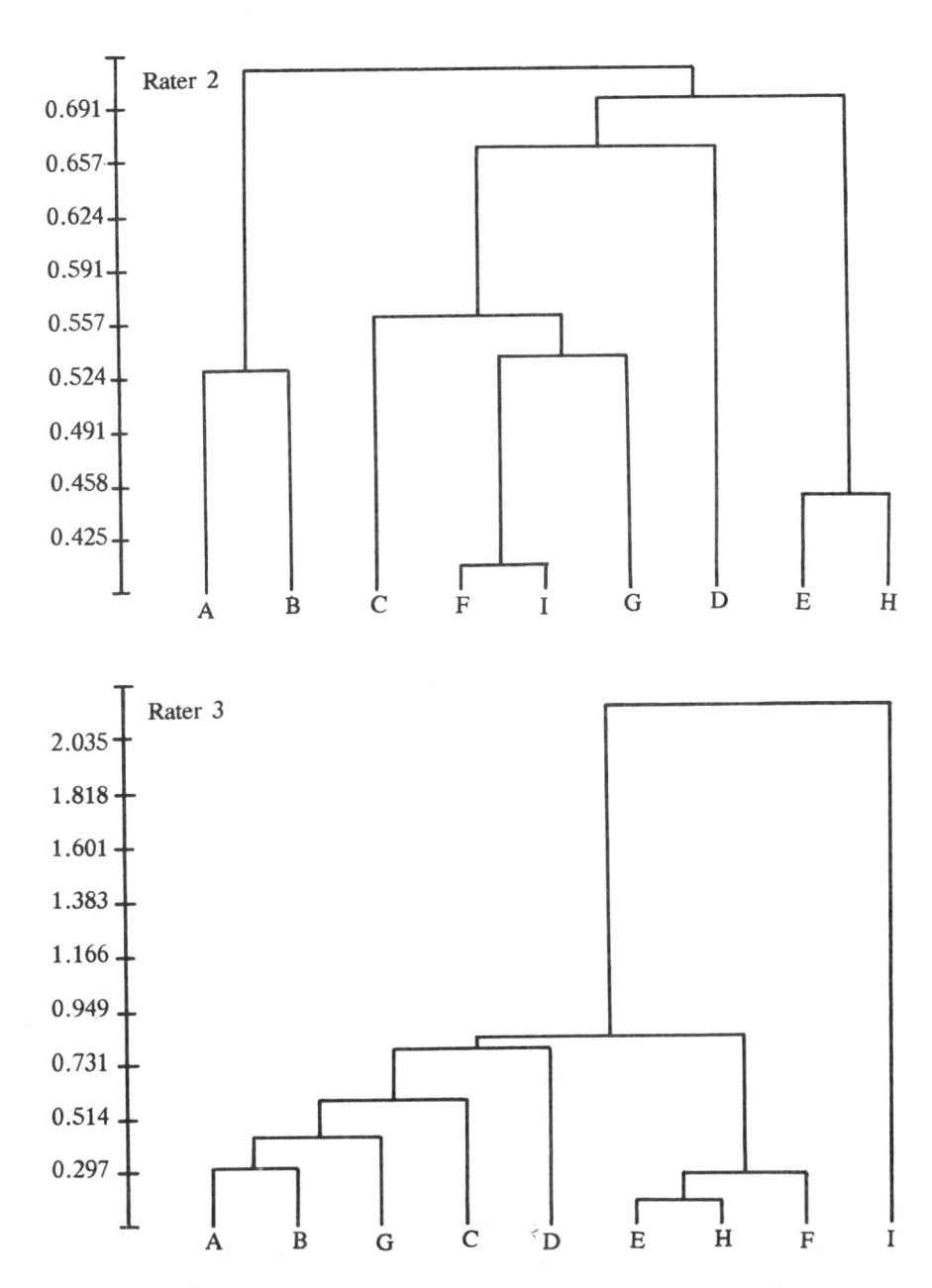

Figure 3.2. continued on next page

Rater 4

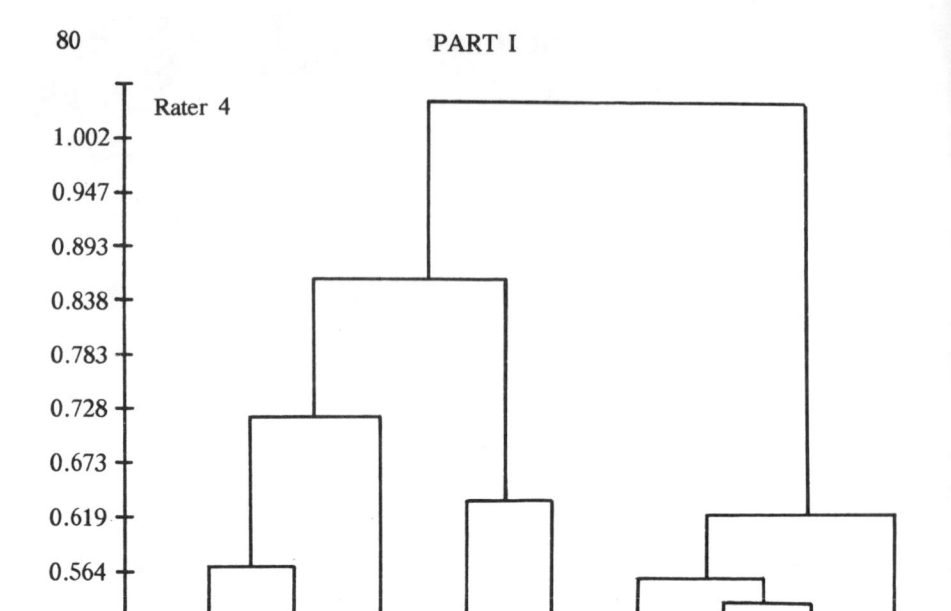

1.002
0.947
0.893
0.838
0.783
0.728
0.673
0.619
0.564

A B G D I C F H E

Figure 3.2.: Dendrograms of intuitiveness — analyticity features (Group Average method) for Rater 1 through Rater 4 (y — axis: minimal values of group variance)

Rater 1: The dendrogram gained with the Group Average method, which is representative for the applied methods (cf. Figure 3.2.), reveals for Rater 1's ratings that the unreliable Feature I is separated from the other features. This is also reflected by correlational analysis that is not reported here. The subclusters (E, F, H) (dependency on personal experience, pictorial metaphors vs. conceptual and numerical/quantitative patterns, and emotional involvement) and (A, B, G) (conscious, pure intellect, high cognitive control) show that Rater 1 is, in some way, distinguishing personal — pictorial — emotional and cognitive aspects of analytic — intuitive thinking.

Rater 2: This rater produces three stable subclusters: (A, B), (E, H), and the extremely homogeneous cluster (F, I), containing Feature I which shows the lowest interrater reliability.

Note that the low level of clustering of these two features in Rater's 2 codings is solely due to an extreme number of nonclassifiable cases. 78 out of 116 subjects could not be classified on Feature F, and 75 on Feature I, resulting in 54 matchings of nonclassifiable subjects. Thus, for Rater 2, like

Rater 1, a quasi−cognitive (A, B) and a personal−emotional cluster (E, H) can be distinguished.

Rater 3: This rater's clustering is similar to that of Rater 1. First, there is the (E, F, H) subcluster. However, in contrast to Rater 1, this subcluster is quasiautomatically generated by the extremely high number of analyticity ratings and the extremely low number of intuitiveness ratings in these features. Furthermore, the subcluster (A, B, G, C) (conscious, pure intellect, sequential step−by−step, cognitive control) may also be identified. In some, but not all cluster methods, the latter subcluster is chained to Feature D (treating the problem structure as a whole). As in Rater 1's rating, the Feature I shows a total outlier characteristic.

Rater 4: Three stable subclusters may be identified for this rater through all cluster methods, namely (E, F, H, C) (dependency on personal experience, pictorial metaphors, emotional involvement, and sudden synthetical processing), the (A, B, G) cluster (cf. Rater 1), and on a more heterogeneous basis, the less reliable Features D and I.

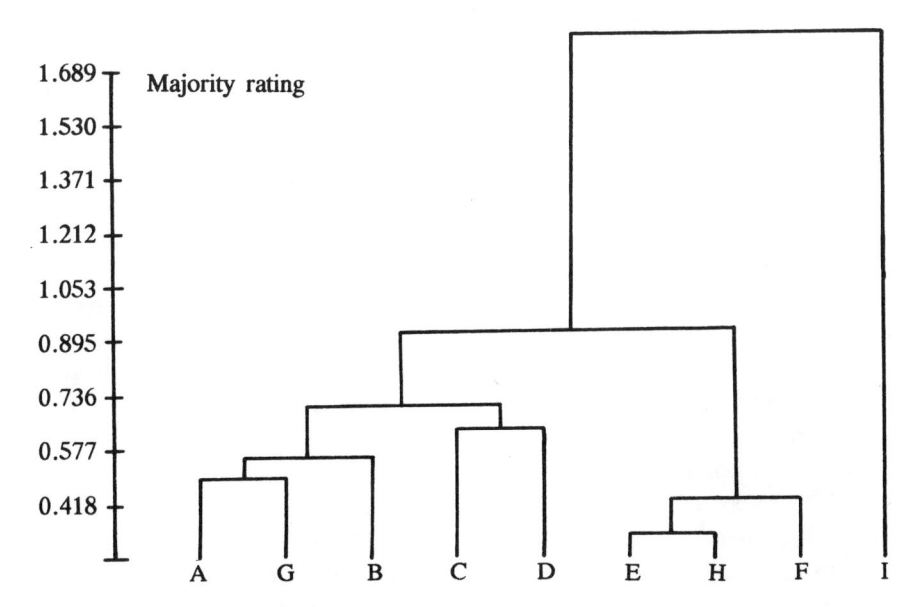

Figure 3.3.: Dendrogram of intuitiveness−analyticity features (Group Average method) for the Majority rating (y−axis: minimal values of group variance)

Majority: The Majority judgment which is based on the mode of the coder's distribution of feature classification (two modes result in the intermediate/ nonclassifiable Majority rating) exactly reproduces Rater 1's stable clusters (A, B, G) and (E, F, H) (cf. Figure 3.3.).

Types of strategies: There are strong similarities in the raters' distributions for the four types of strategies; analytic−algebraic, analytic−nonalgebraic, intuitive−algebraic, and intuitive−nonalgebraic. More than two−thirds of the subjects use a nonalgebraic strategy, about half of these strategies are analytic, and the other half intuitive. Hence there are a lot of analytic−nonalgebraic strategies, but according to the Majority rating, only a minority of two classified strategies which are judged to be algebraic−intuitive (cf. Table 3.7.).

Table 3.7.: Frequency table of algebraic vs. nonalgebraic x analytic vs. intuitive strategies according to the Majority rating

	Mode of thought			
Algebraicity	Intuitive	Analytic	Nonclassifiable	n
Algebraic	2	26	2	30
Nonalgebraic	29	44	7	80
Nonclassifiable	0	6	2	8
n	31	76	11	118

The variety of algebraic approaches which were included in subjects' justifications is particularly interesting, if not overwhelming. As we mentioned above, unfortunately no calculation ended up with the 'normative solution'. The algebraic operations applied are listed in Table 3.8.

Modes of thought and response behavior: The response distributions differ between the modes of thought. Table 3.9. presents the frequency of diagnosticity responses split into the different types of strategies according to the classifications of the individual raters and the Majority classification. According to all ratings, most of the diagnosticity responses are given in the analytic mode. This also holds true for relative frequencies. The difference is significant for all raters and the Majority rating.

Table 3.8.: Algebraic operations produced in the written protocols (as listed by Rater 2 (in brackets: number of cases observed)

A) Simple operations using only one probability or percentage
1. $(1-p(H))-p(H)$ (1)
2. $0.5 \cdot p(H)$ (1)
3. $1-p(D/H)$ (5)
4. $1-p(H)$ (3)

B) Simple connection/linkage of $p(D/H)$ and $p(H)$
1. $p(H) \cdot p(D/H)$ (5)
2. $p(H)+(1-p(D/H))$ (4)
3. $p(H) \cdot (1-p(D/H))$ (1)

C) Means and multiple connections/linkages of $p(D/H)$, $1-p(D/H)$, $p(H)$ and $1-p(H)$
1. mean $(p(D/H), p(H))$ (5)
2. mean $(1-p(H), p(D/H))$ (1)
3. $(1-p(H)) \cdot (1-p(D/H)) \cdot (1-p(H))$ (1)
4. $(1-p(H))-0.5 \cdot (1-p(D/H))$ (1)
5. $p(H)+0.5 \cdot (1-p(D/H))$ (1)
6. $p(H) \cdot (1-p(D/H)) \cdot p(H)$ (1)
7. $p(H)-p(H) \cdot (1-p(D/H)$ (1)
8. $1-((1-p(H)) \cdot (1-p(D/H))+p(H) \cdot p(D/H)$ (1)
9. $p(H) \cdot p(D/H)+(1-p(D/H)) \cdot (1-p(H))$ (1)

D) Further calculations
1. Determination of a ratio of pro and contra arguments (1)

These effects remain significant, even when the nine subjects who started with reproductions of "inverting conditional probability" are excluded. The mean estimation is .087 higher than the 'normative solution' in the analytic mode, but .047 lower than the 'normative solution' in the intuitive mode. (The values are reported for Rater 1's codings.) This difference is nonsignificant. The tendency may be explained by the higher percentage of diagnosticity responses in the analytic mode. Both the absolute value of the deviation $|d_1|$, and a normalized $|d_2|$ deviation (these values are explained in Chapter 2.6.4) do not differ significantly between modes.

Table 3.9.: Frequency of diagnosticity responses split into the different types of strategies according to the raters' classifications (in brackets: total number of strategies according to each rater; the last column presents the Chi^2 value for the comparison between frequencies of analytic and intuitive diagnosticity responses)

	Analytic	Intermediate/ nonclassifiable	Intuitive	Chi^2 df = 1
Rater 1	30 (69)	2 (12)	3 (37)	9.9
Rater 2	31 (75)	3 (7)	1 (35)	16.6
Rater 3	33 (85)	1 (13)	1 (29)	7.0
Rater 4	27 (56)	5 (13)	3 (49)	20.3
Majority	33 (73)	1 (13)	1 (31)	15.6

We will close the results section by reporting the frequency of explicit anchoring and adjustment justifications. The criterion for giving a strategy an anchoring and adjustment rating is that the subject starts with one of the two parameters $p(D|H)$ or $p(H)$, and then modifies it toward the other one. Within the sample of 118 justifications, only 7 strategies were judged by Rater 1 to be of an anchoring and adjustment type.

3.4. SUMMARY AND DISCUSSION OF EXPERIMENT B'S RESULTS

Obviously, a considerable proportion of the subjects show a different understanding of the question in base–rate problems than that intended by the experimenter. Although our subjects were asked to rewrite the question with the original question still present, and about half of the subjects produced copy–like reproductions, nearly 20% started with a reproduction of the inverted probability (which is identical to the diagnosticity in the problems used in our experiment). Other subjects transformed the question into the question on the base–rates or a conjunctive probability, reproduced idiosyncratic reproductions, or noted nothing at all.

There may be at least three reasons why people show this behavior and conduct such transformations. First, as we previously pointed out, verbal formulations of conditional probability are often ambiguous (cf. Chapter 2.2.; BAR–HILLEL, 1984; SCHOLZ & BENTRUP, 1984) and incomplete (cf. SCHOLZ, 1981). This is often the case with letter codified probabilistic tasks (cf. BAR–HILLEL & FALK, 1982). Secondly, subjects may realize that the problem goes beyond their resources, and they may intentionally try to look for a similar problem or question that is less sophisticated. Such behavior has been observed in many problem solving experiments (cf. DÖRNER, 1981; DÖRNER, KREUZIG, REITHER, & STÄUDEL, 1983). Thirdly, the subjects' (mis)understanding may be due to their not possessing the concept of conditional probability or even that they do not possess a probability concept at all. Though individuals are usually able to handle the calculation of conditional frequencies, we know that the concept of conditional probability is much harder to acquire and is obviously not a natural concept which develops spontaneously. Subjects therefore may be overburdened and hence they may, almost of necessity, (unconsciously) misunderstand the question in base–rate problems.

Different modes of thought exist when subjects try to cope with base–rate problems in an experimental setting. We introduced the complementary concepts of intuitive and analytic thinking with lists of nine features. These features may be regarded as dependent subscales of the intuitiveness–analyticity complementarity of thought. The purpose of this definition by lists of features is: (1) to provide a theoretically more substantiated and differentiated description of the modes of thought, and (2) to supply an empirically accessible and reliable procedure for classifying modes of thought. Both the results of the interrater reliabilities produced by four raters and of the study of multivariate contingencies via cluster analysis provided satisfactory results. The cluster analysis on the coding of features produced by all four raters indicates that the modes of thought may be subdivided into cognitive and personal–emotional–motivational aspects (cf. DÖRNER, 1984, p. 18).

Two features, i.e., Feature D (treating the problem structure as a whole vs. separating details of information) and I (feeling of certainty vs. uncertainty toward the product of thinking), showed a completely unsatisfactory interrater reliability. Feature I shows an isolated and idiosyncratic characteristic in the ratings given by Rater 1, Rater 3, and the Majority rating. The results gained with the composite scores of intuitive versus analytic thinking, which were used for the strategy classification, however, were not essentially affected when these features were removed. It has already been theoretically questioned whether a lack of certainty toward the product of the decision process is typical for intuition. We suggest that a next step should be to measure confidence directly by a post hoc questionnaire, which will be performed in another survey (cf. Experiment D), and that one should cancel this feature from the list if similarly poor findings are obtained.

According to our expectations, both analytic—nonalgebraic and intuitive—algebraic strategies exist. There are many nonalgebraic justifications in the analytic mode of thought, and few algebraic justifications assigned to the class of intuitive strategies. In order to illustrate the main characteristics of these four different strategies, we will present prototypes for each of them.

Analytic—algebraic strategies

We begin with a classical analytic—algebraic justification given by a 24—year—old male postgraduate student. He worked on the TV problem with a .02 base—rate for defective screens and a .90 test accuracy, but started from the following reformulation of the question which has been classified as "Using probability when inverting conditional probability" (IIA):

"What is sought is the probability of a red lamp lighting up in the case of a defective tube."

Es soll die Wahrscheinlichkeit genannt werden, daß bei einer defekten Röhre die rote Lampe leuchtet.

Then came the following justifications:

"2% defective screens, this may be subtracted from the total number. The test device gives an incorrect reading for 10% of the remaining 98%, (9.8%); this has to be subtracted again. Hence the result is 88.2%."

2% defekte Röhren diese kann man von der Gesamtzahl abziehen von den verbleibenden 98% zeigt das Prüfgerät 10% fehlerhaft an (9.8%), diese müssen nochmals subtrahiert werden. Das Ergebnis lautet also 88,2%.

We will report Rater 1's coding for this and the subsequent examples, as his ratings are rather representative, though they perhaps tend to feature a relatively high percentage of intermediate/nonclassifiable codings.

He attributed the following analytic features to this strategy: It is a sequential, linear, step—by—step ordered cognitive activity (C), which separates details of information (D), it is independent of personal experience (E), uses (conceptual or) numerical patterns (F), and is generally cold and emotion—free (H). Rater 1 also judged the justification to be confident (I, intuitive), but the other dimensions have not been coded.

An 18-year-old female 12th grader supplied one of the analytic-algebraic justifications resulting in a diagnosticity response. She worked on a TV-problem starting from an "Inverting probability" reformulation and all features of analytic thought were attributed to the subsequent justification.

> "I have added together the tests on the faultless and the defective tubes. The result was 100% (logical). From this 100% I now subtract the 10% of tests with an additional impulse and this yields 90%, which is the probability which arises, if the red light lights up for a tube."
> Ich habe die Messungen der einwandfreien und defekten Röhren zusammengezählt. Das Ergebnis betrug 100% (logisch) Von diesen 100% ziehe ich jetzt die 10% der Messungen bei einem Zusatzimpuls ab und erhalte dabei 90%. Die Wahrscheinlichkeit die auftritt wenn bei einer Röhre eine rote Lampe aufleuchtet.

Intuitive-nonalgebraic strategies

Unequivocal examples of the intuitive-nonalgebraic strategy type are the subsequent two justifications given by a 22-year-old female postgraduate student and a 13-year-old 7th grader when working on the Hit problem.

> "The crucial question in this problem remains whether an audience is influenced by the studio guest's opinion. In most cases, the musical taste of the two (listener and guest) will not be the same, as each of them is committed to his own favorite music. This is why I have set the possibility for the predicted hits at 50:50."
> Die entscheidende Frage bleibt in diesem Problem, ob ein Publikum von der Meinung des Studiogastes beeinflußt wird. In den meisten Fällen wird der Musikgeschmack von beiden (Hörer und Gast) nicht einheitlich sein, da jeder auf seine Lieblingsmusik fixiert ist. Deshalb habe ich die Möglichkeit für die getippten Hits auf 50 : 50 gesetzt.

> "The probability is 30%.
> I believe it must be a small probability, as every human being has a different opinion about music. I also believe that the person who is a guest in the studio must have a neutral attitude to music, for he basically states his own opinion."
> Die Wahrscheinlichkeit beträgt 30%.
> Ich glaube es muß eine geringe Wahrscheinlichkeit sein, weil doch jeder Mensch eine andere Meinung über Musik hat. Ich glaube auch das der Mensch der da Studiogast ist eine neutrale Art zur Musik haben muß denn er vertritt im Grunde ja seine eigene Meinung.

According to Rater 1's codings, the former of these two justifications, for instance, may be sketched as follows: The subjects' argumentation and understanding is influenced by feeling (B), obviously subjective past experience with musical taste (E) is essential, and various background experiences result in a sudden synthetical assessment (C) of 50:50. The problem seems to be treated as a whole (D), and the pronounced formulations presumably reveal a considerable emotional involvement (H). As the subject's arguments appear to be (subjectively) reasoned and conscious (A, analytic), 5 intuitive and 1 analytic features are attributed.

Intuitive — algebraic strategies

We will present two similar justifications. The first was given by a 14 — year — old male 7th grade student, the second by a 22 — year — old female postgraduate student. Both worked on the Motor problem with a diagnosticity of .80; the former's base — rate, however, was .3 whereas the latter's was .05.

"I proceeded by calculation. I have calculated 3/10 + 8/10, but instead of keeping the same denominator, I have also added the two denominators; result 11/20. So, in my opinion, I have added all chances together to end up with one chance. It is difficult for me to explain why I have chosen this path."

Ich bin rechnerisch vorgegangen, ich habe 3/10 + 8/10 errechnet, aber statt denselben Nenner zu behalten, addierte ich auch die Nenner; ergebnis 11/20.
So habe ich meiner Meinung nach alle Chancen zusammen addiert, um eine Chance zu haben. Es ist für mich schwer zu erklären, wieso ich diesen Weg genommen habe.

"87,5
95% + 80% = 175 : 2 = 87.5%
Why this calculation is not quite clear to me, but it seems to me to be intuitively correct."

87,5
95% + 80% = 175 : 2 = 87,5%
Wieso diese Rechnung, ist mir auch nicht ganz klar, erscheint mir eher intuitiv richtig.

We will sketch Rater 1's coding of the first example. The strategy was judged to be of preconscious information processing (A), accompanied by understanding by feeling and instinct of empathy (B). The problem structure was treated as a whole (D), as one formula ties all information, and there is a low cognitive control on the process (G). In contrast, however, there is a clear independence from personal past experience (E, analytic). The other features have not been classified. In the second example all features except F were assigned the label intuitive. The Feature F was rated on the analytic side, as numerical patterns are clearly dominant.

Analytic — nonalgebraic strategies.

We will conclude the series of prototypes, and the demonstration of the applicability of the analyticity — intuitiveness rating procedure, by presenting a justification given by a 24 year old male postgraduate student. When working on the Hit problem he wrote:

"Answer: 70%
According to the data, 4 out of 5 titles become hits were also judged as hits by the studio guest; and that over the previous five years. A title selected at random thus has an 80% chance of getting into the hit parade, if it has been named a hit by the studio guest. I have reduced this by 10%, as only 35% of the titles actually got into the hit parade."

Antwort 70%
Der Statistik nach werden 4 von 5 Schlagern Hits, die auch vom Studiogast als Hit beurteilt werden und das in den letzten 5 Jahren. Ein zufällig ausgewählter Schlager hat also die Chance zu 80% in die Hitparade zu kommen, wenn er von Studiogast als Hit bezeichnet wird. 10% werden von mir deswegen abgezogen weil nur 35% der vorgestellten Titel in die Hitparade kamen.

This procedure seems to be conscious (A) and logical (B), it is a sequential activity starting from 80%, then lowering this probability in another step (C). There are no arguments introduced which are taken from personal experience (E), and the operations are controlled, justified (G), and cold (H). Unclear, however, is confidence and also Feature D, which may be rated on the intuitive pole as all informations are tied. Nevertheless, there is a clear majority of analytic features.

As reported above, all justifications were given after a second trial. On the first trial no justifications were requested, but some of the subjects were asked to give a weighting of the significance of both base — rate information and diagnosticity information. The responses were given on a 6 — point scale with the poles "not at all" (1) and "very significant" (6). Consistent with his justification, the above subject assigned a weight of five to the diagnosticity and one to the base — rate information.

3.5.   CONCLUSIONS

The content analysis of the subjects' reformulations of questions and written justifications substantiates the necessity of starting from the individual's text understanding and analyzing the judgmental process if one wishes to investigate probability judgments. This would appear to be self-evident, but hitherto, hardly any experimental study in our field actually commenced with an analysis of subjects' problem understanding.

When analyzing the 118 written justifications, only slight evidence could be explicitly found indicating that the widely discussed **heuristics** like the representativeness heuristic, the causal schema, or relevance-specificity considerations are guiding the decision process of more than a small number of subjects. On the other hand, at least 9 out of the 34 subjects who responded with the diagnosticity started from a reformulation of the inverted probability. Hence these heuristics can only account for some of the diagnosticity responses. Representativeness is only one possible rationale for the diagnosticity response. As has been pointed out in Chapter 2.3., drawing conclusions about the cognitive process from the response, as is often attempted in decision research, is not legitimate. Similar findings are reported by POLLATSEK, KONOLD, WELL, and LIMA, 1984, where considerably less than half of the sample of subjects, who gave a representativeness response in a random sampling experiment, supplied arguments that could be interpreted with the representativeness heuristic.

One conventional explanation for intermediate responses in base-rate problems is the **anchoring and adjustment** hypothesis. This hypothesis states that subjects are anchoring on some data (e.g., on the diagnosticity information), and then show an insufficient adjustment toward the correct or adequate solution (NISBETT & ROSS, 1980, p. 41). Anchoring and adjustment is a strategy that occurs in base-rate problems. In variants of the Cab problem, it may be either of an **analytic-nonalgebraic type** (see the example given above), or an **intuitive-nonalgebraic type**. If a nonrandom probability judgment lies between the base-rate and the diagnosticity, an analytic acquisition of and penetration into the problem structure seems to be necessary. Further, an understanding of the interaction of base-rate and diagnosticity information is required. Whether the specific adjustment procedure following the (analytic) structuring phase is of the analytic or intuitive mode, or whether this process is a conscious one, cannot generally be answered.

An oscillation to and fro between the modes of thought seems to be desirable. Various unwanted disadvantages of the two modes of thought may be prevented if the modes are properly interlocked. The fact, however, that only relatively few protocols were judged to carry more than one feature of both modes may indicate that pure analytic or pure intuitive strategies are predominant in our experiment. A pure **analytic-algebraic** strategy may lead to a precise determination of the 'normative solution'. Within the **analytic—algebraic** strategies, a multitude of different algebraic operations are applied. However, there is no doubt that these algebraic operations did not help our subjects to achieve the aim of determining a precise and adequate probability. On the contrary, many of the calculations resulted in crude and unsophisticated responses, mainly in the analytic mode. Often the calculations performed seem to be the product of a blind, insightless, and mechanistic process, as if a conditioned response pattern is elicited, roughly taking the form: **probability — mathematics — pick up the numbers — produce a calculation — perhaps it fits**.

The main difference between the **intuitive—algebraic** and the **analytic—algebraic** strategies in Experiment B is that the former are (mostly) a sudden understanding by feeling which treat the problem structure as a whole, whereas the latter are characterized by a quasi—mechanical, sequential, step—by—step, separating, cold, emotion—free treatment of the information. **Intuitive—nonalgebraic** strategies sometimes make a rough, fuzzy, cursory, vague, and sketchy impression, and sometimes personal experience is also introduced which does not seem to meet the problem. The latter is especially the case if the problem is not taken from the subject's world, is artificial, or the subject's perception does not meet the problem.

If we want to sum up what is gained by the introduction of the above classification of strategies, three issues are substantial. **First**, a description and definition of essentially different cognitive activities and processes which are responsible for probability judgments has been provided. **Second**, this classification of cognitive strategies may be applied in a rating procedure and yields a satisfactory interrater reliability, as shown by the above four rater study. **Third**, this description may supply foundations for a more precise conceptualization of heuristics in decision making under uncertainty and the general process of stochastic thinking. Thus, the following could be specified: **Anchoring and adjustment** or information weighting with subsequent information integration may be present in both modes of thought, but seems to be relevant mainly among older, more highly educated subjects (cf. Chapter 2.8.). **Wild calculations**, analytic but noninsightful and unsophisticated algebraic operations (cf. Table 3.8.) were a common analytic procedure used by our subjects. Furthermore, there is evidence of intuitive reasoning that does not refer to the "classical heuristics of probability judgments" (see above). Only slight knowledge could be gained about the classical heuristics (e.g., the representativeness heuristic) which have been supposed to be responsible for (most of) the diagnosticity responses. These heuristics have been called **intuitive predictions** (cf. KAHNEMAN & TVERSKY, 1973, p. 237; TVERSKY & KAHNEMAN, 1974; 1982, p. 494). However, there is a **fourth** substantial finding that is not only worth mentioning but that also might have important theoretical implications. Contrary to the predominant view (cf. KAHNEMAN & TVERSKY, 1973; TVERSKY & KAHNEMAN, 1979), under the above experimental conditions, diagnosticity responses were mostly produced in the analytic mode and were not (fallacious) intuitive responses. Sometimes diagnosticity responses are also obtained by explicit calculation, (cf. the second example of an analytic—algebraic strategy). How and whether the classical heuristics may be tied to the modes of thought, and whether they are really guiding nearly half of the subjects' responses, as has been suggested by previous research (cf. Chapter 2.3.), or whether the results of this study may be replicated, will be investigated in Experiment C (next Chapter).

4. MODES OF THOUGHT AND PROBLEM FRAMING IN THE STOCHASTIC THINKING OF STUDENTS AND EXPERTS (SOPHISTICATED DECISION MAKERS)

In the last chapter we introduced a conceptualization of two substantially different modes of thought in persons coping with probabilistic problems and probability judgments. The modes of thought, which are considered to be crucial for both a more differentiated understanding and for the development of a model of stochastic thinking, are the complementary modes of **analytic** versus **intuitive** thought.

The characteristics and features of these two different types of processing have been defined by a list of features or attributes (see Table 3.1; a description of the individual features is given in Chapter 3.2.). In an experiment using several variants of KAHNEMAN and TVERSKY's, 1973, Cab problem, short written justifications of subjects' probability judgments were rated according to these features. A subject was judged to have processed in an analytic/intuitive manner if a definite majority of features (at least two more of one than the other) could be assigned to one of these modes.

4.1. THE CONCERNS OF THIS CHAPTER

Improving the procedure for tracing a subject's cognitive process: The experimental procedure of Experiment B (cf. Chapter 3.3.) had some unwanted properties. First, subjects' text understanding could only be controlled roughly with brief written reformulations of the questions on the requested probability. Second, the amount of time subjects could spend on the problem was very limited, as the subjects had to digest the instructions and deal with five problems within one hour. Of course, an analysis of rapid estimations and quick judgments following the text reception or problem acquisition are also of interest. Many judgments and decisions in everyday life have to be produced under time pressure, and, from a theoretical perspective, the starting point or initial phase of a cognitive process is as interesting as the terminal phase. But PHILLIPS', 1983, critique of decision research as a psychology of first impression may also be interpreted as a request for a tracing of subjects' penetration into a task that leads to a sophisticated and (subjectively) well−grounded response. One of the aims of the forthcoming Experiment C was to carefully control subjects' understanding of text and question, and to trace the course taken by the subjects during iterated working on one and the same problem, until they have settled on a well−considered and reflected response, which presumably will not be rapidly revised. In order to do this, we designed an experimental procedure which permits the control of subjects' problem understanding, and provides information about their cognitive strategies. Furthermore we employed another rough control over the appropriateness of the written justifications by directly questioning the subjects. This procedure was not included in Experiment B.

Testing hypotheses about the dependency of the modes of thought: The construct of analytic and intuitive thought in stochastic thinking was introduced by the lists of features when trying to evaluate previously produced written justifications. Hence, hypotheses about the modes of thought could not be tested in this previous experiment. In this chapter, several hypotheses are introduced that concern the dependency of the mode of thought on situational and differential variables (cf. Chapter 4.2.). These hypotheses are tested in Experiment C. In another Experiment D, an analysis is performed in order to find out whether the modes of thought exist, and which responses are produced by the different modi in a group of professional experts in stochastic. The theoretical questions and expectations underlying this expert study are dealt with in Chapter 4.6. Both experiments may be considered as classical construct validations (cf. APA, 1954; WEINER, 1976, p. 14) of the introduced method of determining the modes of thought by a rating procedure based on the proposed lists of attributes and features.

Heuristics, modes of thought, and the 'base−rate fallacy': The analysis of Experiment B (cf. Chapter 2) produced some surprising post hoc findings on how the diagnosticity responses and judgmental heuristics are tied to the modes of thought. In the forthcoming experiments, we will once more investigate whether the diagnosticity responses are typical analytic responses or not, as we consider this question to be of considerable theoretical importance.

4.2. THE IMPACT OF PROBLEM FRAMING, ITERATION, SEX, AND CAREER SOCIALIZATION ON THE ACTIVATED MODE OF THOUGHT

One of our basic assumptions is that the specific mode of thought which is applied and activated is fixed by situational (cf. UNGSON & BRAUNSTEIN, 1982, p. 4) and differential variables. It is assumed that the mode of thought activated varies intra— and interpersonally.

In order to describe essential situational variables that could influence the way a subject copes with stochastic situations, we will use the concept of **problem framing**. The term 'framing' or 'frame' is currently widely used, and different notions are assigned to these concepts. Within this volume, we will not only introduce the concept problem framing but also cognitive framing (in the next chapter). We should note that at least the notion of problem framing deviates from the common use of the framing concept. Prior to specifying the meaning of problem framing, we will add some general remarks on the frame concept in order to illustrate not only the range of its use but also the general denominator of the diverse applications. The frame concept has been independently developed by the artificial intelligence researcher MINSKY, 1975, and the sociologist GOFFMAN, 1974. In both approaches, it is employed to conceptualize the effect that the context, meaning, or content of a problem has on the cognitive activity.

KAHNEMAN and TVERSKY were the first to introduce the frame concept into decision research. Within their prospect theory (cf. KAHNEMAN & TVERSKY, 1979c), they outlined that the subjective evaluation of the difference between two outcomes depends on the reference point. If, for instance, one has to choose between two bets with different outcomes, a change of the reference point may imply a change of preference, although the bets (i.e., the outcomes and their probability of occurrence) remain the same (cf. TVERSKY & KAHNEMAN, 1981). According to KAHNEMAN and TVERSKY it matters whether an outcome is compared with a better one (and hence regarded as a loss) or whether it is compared with a worse one (and thus viewed as a gain). Several examples that show how task features and contextual considerations affect the framing of judgment, the reference point, and hence the response behavior have been described by FISCHHOFF, 1983b; SLOVIC, FISCHHOFF, and LICHTENSTEIN, 1982; SCHOEMAKER and KUNREUTHER, 1979; SLOVIC, 1983; HERSHEY and SCHOEMAKER, 1980.

The following effects of content on the individual's understanding may also be regarded as framing effects. Both subjects' understanding of logical relations (cf. WASON & SHAPIRO, 1971; JOHNSON–LAIRD & WASON, 1977; COX & GRIGGS, 1982), and also their sensitivity to and understanding of conditional probabilities, obviously also depend on the **content of the problem**. In Chapter 3.1. (Section 2), we have already mentioned the difficulties involved in the subjects' understanding of the question on the conditional probability in the so—called Tom W. problem (cf. Chapter 2.1.). BAR–HILLEL's, 1984, findings suggest that subjects have extreme difficulty in discriminating between the conditional probabilities $p(H_i|D)$ and $p(D|H_i)$ (i.e., the probability that Tom W. is working in the field H_i, given the personality description D, from the probability that a description D has been produced for some person, given that he is working in the field H_i). However, a change in cover (framing) presumably causes all difficulties to disappear. Hardly anyone will fail to distinguish, for instance, the probability $p(W|P)$ that someone is white who has been randomly chosen from the sample of all popes who have ever lived (which, as far as we know is a probability of 1.0), from the inverted probability of $p(P|W)$, i.e., that someone is the pope given the information that he is white, which probability is near to zero.

In her review on the 'base—rate fallacy', BAR–HILLEL, 1983, discriminates between a textbook paradigm and a social judgment paradigm of the 'base—rate fallacy'. An example of the former is the Cab problem, a prototype of the latter, the Tom W. problem (cf. KAHNEMAN & TVERSKY, 1973). According to BAR–HILLEL, the main difference between these two paradigms is that within the latter, the judgments are based on the representativeness heuristic, whereas in the different variants of the Cab problem, other strategies are employed. We think, however, that the Cab paradigm of the 'base—rate fallacy' may be differently framed, and this may lead to the appearance of the different conceptualizations and cognitive strategies described by BAR–HILLEL in one and the same paradigm of the 'base—rate fallacy'.

The following distinction between a **problem solving framing** and a **social judgment framing** of a problem structure refers to the discrimination made by BAR–HILLEL and the general frame concept quoted above. However, we intend to broaden the meaning of the frame concept as both types of the previous framings may not only be specified by the **internal problem properties or the content of a problem,** but also by the **circumstances** under which the problem is presented to an individual. We introduce the construct of problem solving versus social judgment framing, as we suppose that both

CHAPTER 4.2.

framing variants may elicit completely different cognitive activities in one and the same problem structure. We will now give general descriptions of the problem and social judgment framing.

The task properties of a prototypical **problem solving framing** are a letter codified form with a rather mathematical textbook image. This implies that the information has to be taken "**as if**", both in the case of a non−real−world artificial story, or in the case of an eventually realistic story. Realistic covers of problem solving frames are mainly taken from the technical world in which the information appears to be precisely and objectively assessed. Hence, the parameters leave little room for interpretation and are thus reliable. There are usually only a few sequentially introduced cues (cf. HAMMOND, HAMM, GRASSIA, & PEARSON, 1983) embedded into a closed story, which scarcely permits the introduction of additional information that may change the problem structure. Consequently, one can usually agree upon a single, unique, 'normative solution' for these problems. Clearly this 'normative solution' is not absolute in the sense of a secundum non datur. In this we follow conventional practice by using the term 'normative' to describe the use of a rule when there is a consensus among formal scientists that one certain rule is appropriate for the particular problem.

In a problem solving framing, the task is presented in a rather formal task−oriented manner. One is expected to produce an exact answer while spending some time on the task.

In a **social judgment framing**, the content appears to be realistic and is mostly taken from the nontechnical world. The information does not seem to be completely precise but more interpretable. There are usually a high number of sometimes concurrent, redundant cues (cf. HAMMOND, 1984), such as, for instance, perceptual cues (cf. PETERS, HAMMOND, & SUMMERS, 1974). These cues, however, may be questioned and evaluated by personal experience, which normally results in there being more than one adequate answer.

If individuals are confronted with a task in a social judgment framing, they are not expected to produce an exact answer, but rather a (well−assessed) opinion or estimation. Mostly, the individual will be expected to assess this estimation based on his personal expertise and experience within a relatively short period of time.

As mentioned above, we hypothesize a contingency between the modes of thought and the framing of a task. It is supposed that individuals will more frequently process in an analytic mode within a problem solving framing compared to a social judgment framing. We know that the attempts to attain an adequate analytic answer, for example the 'normative solution', are often unsuccessful. Furthermore, if algebraic methods are applied, which is often the case in an analytic mode that is supposed to be linked to a problem solving framing, extreme errors in the case of the base−rate problems under consideration (especially unreasonably low probabilities) are expected under the conditions of a problem solving framing (cf. HAMMOND, 1984; HAMMOND, HAMM, GRASSIA, & PEARSON, 1983).

The second variable to be investigated is the **iterated working** on one and the same problem. We designed an experimental procedure which allows a thorough analysis of the subject's problem understanding and response strategies in three rounds of coping with a base−rate problem. In the results section, we will start by analyzing the written reproductions of the complete problem, including the question on the requested conditional probability, after the subject has undergone a careful and intensive process of problem acquisition, and will analyze a first response using the same method that was used in the previous experiments on the 'base−rate fallacy'. To check how stable these responses are in the short run, and whether there are superficial errors in subjects' problem acquisition, the subjects were asked to repeat the complete procedure of the first round in another trial. In a further step, subjects were asked to deal with the problem over a longer time period, i.e., one or two days. The purpose of this was to educe the subjects' best and most sophisticated answer.

Various investigations have repeatedly shown that iterated learning, longer extended working on, or discussion of base−rate problems induces only a slight modification in the response behavior, especially if no feedback on the response is given (cf. LYON & SLOVIC, 1976; FISCHHOFF, SLOVIC, & LICHTENSTEIN, 1979; or the findings of Chapter 2.). Although no significant shifts are thus to be expected with regard to the first two trials, in the third trial, in which the subjects were sent home, a change in the response behavior may be expected. Outside the laboratory the subjects might be freed from the stress of the experimental procedure, they might delve deeper into the problem structure, and more cues may show up in the protocols. Eventually, external resources, like books on probability, may also be consulted, or through discussions with other persons on the problem, the inadequacy of a first glance response may be realized.

Furthermore, we expect some changes in the mode of thought. As the subjects are compelled to cope with the problems three times, a more analytic treatment should result. We thus expect an increase in analytic features, more 'normative solutions', and fewer diagnosticity responses. Interactions with the framing variable and the other variables are investigated in an exploratory manner.

The sex variable has been rather controversially discussed in the literature on styles of thinking and achievement in tasks related to mathematics. The ability to think analytically is often considered to be a personal trait, and also to be dependent on sex. Studies on stochastic thinking have also produced some findings that might support sex dependency in the use of the modes of thought. GREEN, 1982a, for instance, reports that in a probability concepts test, boys performed better than girls in their total scores. However — as he states — some "results suggest that girls are intuitively better" (p. 99). Developmental studies on the acquisition of the probability concept have also produced inconsistent results (cf. ROSS, 1966; CHAPMAN, 1975; PERNER, 1979; HOEMANN & ROSS, 1982). The slight trend in favor of males in some experiments has to be discussed cautiously, as male and female subjects have never been matched for cognitive level prior to the experiment (cf. HOEMANN & ROSS, 1982). As we will see, this deficiency may also be found in other fields of gender research which are relevant to our investigation.

The sex dependency of intuition sometimes seems to be an almost common belief which is still put forward by some papers on intuition. For instance, BASTICK, 1982, p. 329, states, "that the constituent abilities necessary for intuition were sex—related". What is meant by this, is unclear. We think, however, that a lesson may be learned from the discussion about sex differences in mathematical achievement. Here, from a phenomenological point of view, the variable sex seems to be causally linked to an individual's ability to perform mathematics. Thus, it is well—known that there are more male professional mathematicians than female, that in national mathematical competitions, for instance, there are a majority of male winners, and presumably a random sample of female world citizens would not perform so well in conventional mathematics tests as a male random sample (cf. SCHILDKAMP—KÜNDIGER, 1980, 1982). However, a closer analysis conveys a much more differentiated view. Large scale studies (cf. HUSEN, 1967; SCHOLZ, 1983d; ROACH, 1979) have, on the contrary, repeatedly shown that the sex factor is, at most, marginally relevant . In causal analyses that include other relevant factors (such as social status, length of instruction in mathematics, etc., cf. ROSIER, 1980; FENNEMA, 1979; or KLIEME,

1986) the sex variable is often irrelevant, indicating that sex itself is not, per se, responsible for this ability. Nevertheless, we controlled the factor Sex in this experiment, while keeping the educational background roughly constant.

Unlike the Sex factor, the factor Career Socialization is assumed to affect the response behavior and to be highly contingent to the modes of thought. Mathematicians and natural scientists are educated to think in formal (sequentially ordered) models and to think analytically. Analysis is an indispensible feature of every science, but within mathematics, physics, and theoretical chemistry it is the absolutely predominant (and the only accepted) way of gaining and communicating knowledge, whereas one may find heuristic interpretative methods in the social sciences. For the factor **Career Socialization**, with the levels Social Science and Natural Science, we thus expect different degrees or intensities of analysis in the response justifications.

4.3. MODES OF THOUGHT AND THE 'BASE–RATE FALLACY' – CRIPPLED INTUITION VERSUS FALLACIOUS ANALYSIS?

Intuitive heuristics have been regarded as the source of the neglect of base–rates (cf. KAHNEMAN & TVERSKY, 1973; TVERSKY & KAHNEMAN, 1979) and thus the cause of the high percentage of diagnosticity responses. Unexpectedly, however, in the above Experiment B, an overwhelming majority of diagnosticity responses were judged to have been produced in an analytic mode, and only rare diagnosticity responses were produced in the intuitive mode of thought. Hardly any evidence could be gained from the justifications to support the hypothesis that the frequently discussed judgmental heuristics (cf. KAHNEMAN, SLOVIC, & TVERSKY, 1982; SCHOLZ, 1981; BAR–HILLEL, 1983; WALLSTEN, 1983) are the cause of these responses. Due to the experimental constraints of Experiment B, only short justifications were produced. Consequently, one of the objectives of this paper will be to scan subjects' justifications more carefully and subtly for indicators of the heuristics that have been applied.

The relation between local and global cognitive activities in stochastic thinking needs to be clarified. On the one hand, heuristics are considered to be general, universal cognitive operations which are applied in different situations (cf. EINHORN, 1980; SCHOLZ, 1981), while on the other hand, they have been formulated in isolation from one another. KRUGLANSKI, 1982, points out that the theoretical plan behind this cognitive pluralism is the microprocess x situational matrix model. If a situation has been analyzed, this model aims to predict which microprocess (e.g., heuristic or decision rule) is applied. A general cognitive model explaining why and when specific inferences are made and which knowledge base is guiding these microprocesses still needs to be elaborated.

The framework of the analytic and intuitive mode of thought, as it has been introduced by the lists and features and as it will be validated in this chapter, may help to provide a better understanding of the nature of the judgmental process in stochastic thinking, but it does not supply a framework for a conceptualization of the **phases** of the judgmental process. A process model of judgment and decision making in stochastic thinking that refers to the findings of Experiments A to D will be introduced in Chapter 5 (Part II of this volume).

4.4. HYPOTHESES

We will summarize the hypotheses which were established in Chapter 4.2. and 4.3., and that are to be tested in Experiment C.

Hypothesis 1: A Social Judgment Problem Frame more frequently elicits features of intuitive thinking than a Problem Solving Frame. Vice versa, a Problem Solving Frame more frequently elicits an analytic mode of thought than a Social Judgment Frame.

With regard to the behavioral data, we anticipate more extreme errors in the Problem Solving Frame.

Hypothesis 2: Iterated treatment of one and the same base−rate problem induces a penetration into the problem which is reflected by an increasing number of analytic features and cues in the protocols, more 'normative solutions', and fewer diagnosticity responses.

Hypothesis 3: More analytic strategies are to be found among students with a Natural Science socialization than among students with a Career Socialization gained in the Social Sciences.

Alongside the factor Career Socialization, the factor Sex is controlled but no effects are expected. Interaction between tne variables is investigated in an exploratory manner. With respect to the diagnosticity responses, contradictory opinions can be found. According to KAHNEMAN and TVERSKY (see above), more diagnosticity responses should be generated in a Social Judgment frame. On the contrary, our results have indicated that diagnosticity responses were almost exclusively generated in an analytic mode. Hence, with respect to the origin of diagnosticity, our study may be considered to be an experimentum crucis.

Hypothesis 4: More diagnosticity responses are generated by the analytic mode than by the intuitive mode of thought.

4.5. EXPERIMENT C

4.5.1. SUBJECTS

The subjects were 32 postgraduate students of Bielefeld University who had received their Vordiplom (a pre-diploma received after two years of university studies). Their mean age was 24.4 years.

4.5.2. PROCEDURE AND EXPERIMENTAL TASKS

The subjects replied to advertisements posted up in all departments of Bielefeld University. Those who were interested in participating in an experiment on information processing or in an experiment on problem solving were asked to call a certain telephone number. Every second male and female subject was told that they could participate in a problem solving experiment in the Institut für Didaktik der Mathematik, the other half of the students were told that they could participate in an experiment on text processing in the Psychology Department. Subjects were paid DM 40.− for their participation.

Due to the objectives of inducing subjects' intensive occupation with the experimental tasks and of gaining detailed protocols on the thought process, a rather painstaking and sophisticated experimental procedure was designed. The instructions differed according to the location of the experiment. In order to facilitate an understanding of this procedure, we will first roughly sketch the course of the whole experiment as it took place in the Psychology Department.

We will then describe the differences in the procedure at the Institut für Didaktik der Mathematik and finally present the wording of the texts/problems.

In the Psychology Department, subjects first determined a random number on a word processor. Then, the subjects were once again informed that they were taking part in an experiment on text processing that was designed to investigate the estimation of percentages and probabilities. They were told that the aim of the first session was to understand how subjects' estimations and probability judgments are generated. In order to achieve this, they should give an extensive written report of their considerations and the justifications for their estimations or probability judgments for each of the three problems presented.

Each text/problem, and the questions on the text/problem, were first introduced on a tape recorder. Subjects were not allowed to make notes while listening. Afterwards, they had to read the text and the question sequentially, sentence by sentence, on a word processor. They were then allowed to write down **notes** on a prepared sheet, No. 1, but should avoid copying the text. The subjects could control the length of time that the sentences were displayed. After they had finished reading each text and the questions on it, subjects had to reproduce and write down the **text** and the **questions** on sheet No. 2, had to **answer** the questions, report their **degree of confidence**, and detail their **justifications** and the **ideas** that came to them while determining their answers on sheets No. 3 and 4. The subjects participating in the Psychology Department were first presented with the Eye Color story (see below), then with modifications of either the TV problem or the Hit Parade problem. The third story was in each case a repetition of the second one. In order to motivate subjects to attack the base−rate problem in yet another trial, they were told that this had been a long and complicated text, and perhaps not everything had been checked immediately on the first presentation, or maybe some information had been overseen. The experimental procedure was designed to prevent the subjects suspecting that the base−rate problem had been repeated because their own first response had been inadequate.

After finishing the reading, reproduction, and justification procedures on a base−rate problem for a second time, subjects were given their homework. They received a text of the last problem and were asked to work on it until the next session (which was either on the next day or two days later), to formulate another well−considered probability judgment, and to protocol their considerations. As one usually cannot prevent subjects from communicating with friends about their tasks or referring to books, they were asked to fill out a questionnaire about any assistance with their homework received from external sources.

Experimental sessions in the Institut für Didaktik der Mathematik differed in the following points. The experiment was called a **problem solving experiment**. Instead of **texts**, the wording **tasks** or **problem tasks** was used, instead of **estimations and judgments**, **solutions or judgments** were required, and instead of the **Eye Color** another text/problem called the **Roulette problem** was presented before either the Hit Parade or TV problem.

The second session included a questionnaire on the time, intensity, involvement, etc. given to the experiment, and a test on probability including tasks on the general addition theorem, stochastic independence, and the formula for conditional probability.

We will now turn to the experimental tasks. The **Roulette problem** and the **Eye Color story** are prototypical examples of a Problem Solving Cover (Roulette problem) and a Social Judgment Cover (Eye Color story).

Roulette problem:
> You are given the chance to participate in one of the following two roulette games.
> In Game A you will gain 1,– DM if the spinner comes to rest when pointing at a black sector in the first trial. Exactly 50% of the area is black.
> In Game B you will gain 1,– DM if the spinner comes to rest when pointing at a black sector in the first, in the second, or in both trials. Exactly 25% of the area is black.
> Question: Which of the following statements is correct?
> – Game A offers a greater chance of winning than Game B.
> – Both games offer the same chance of winning.
> – Game B offers a greater chance of winning than Game A.

Eye Color story:

As mentioned above, subjects who participated in the Psychology Department were welcomed by the experimenter and then asked to participate in a brief random experiment. They were instructed to press a key on a word processor to stop a programmed counting procedure. The random number at which the key had been pressed was displayed on the computer screen. Then both the instructions and the following text of the Eye Color story were played back on the tape recorder.

Each West German citizen is obliged to state the color of his eyes in his identity card.
Please correct the following statement.
Please notice that you have determined the percentage value inserted in the following statement during the preceding random experiment. By no means can this number provide an orientation for your estimation.
The statement: 11% of all West German citizens have written in their identity card: "Eyecolor: blue".
Question: Instead of 11%, what percentage is correct?

The **base–rate** problems were the core of this paper and are adapted from Experiment A. The **Hit Parade** problem was slightly modified (cf. Chapter 2.1.).

The statement,
The studio guests made 80% correct predictions, both for the songs that became accepted into a hit parade and for those that did not

was refined in the following way, so that the meanings of correct predictions in the case of an actual hit or non–hit were given:
He (the fan who recorded the data) also found out that the studio guests had made 80% correct predictions, both for the titles that became hits and the titles that were rejected. In other words, the studio guest made a correct hit–prediction for 80% of the titles that actually became hits, and a correct no–hit–prediction for 80% of the titles that did not become hits.

The **TV problem** was almost identical to the TV problem presented in Chapter 2.1. In order to prevent interpretations like the one sketched in Example 2, Chapter 2.2., the second paragraph of the TV problem:
Tests over several years have shown that 75% of the tubes are okay and 25% are defective.

was replaced with the following text:
In reality the following is true: 75% of the tubes are okay and 25% are defective.

4.5.3. INDEPENDENT VARIABLES

There are two framing operationalizations or variants; an **Experimental Framing** and a **Cover or Content Framing** of the base–rate problems. Whereas the latter framing variant is defined by internal text/problem properties, the Experimental Framing refers to the circumstances and the procedure under which the base–rate problems are exposed to the subjects.
We distinguish between an Experimental Social Judgment Frame, given by the procedure in the Psychology Department, and an Experimental Problem

Solving Frame, established by the procedure in the Institut für Didaktik der Mathematik. The Hit Parade problem is called a Social Judgment Cover Frame, and the TV problem, a Problem Solving Cover Frame for the base—rate stories (see Table 4.1).

Table 4.1.: Variants of framing manipulated in the experiment, number of subjects inserted for each cell

		Experimental Framing	
		Social Judgment	Problem Solving
Cover Framing	Social Judgment	8	8
	Problem Solving	8	8

Half of the subjects per cell were Social Science students and half Natural Science students (factor **Career Socialization**). Furthermore, half of the subjects were male and half female (factor **Sex**). The factors Experimental Framing, Cover Framing, Career Socialization, and Sex constitute a complete 2x2x2x2 factorial experimental design with two subjects per cell. With regard to the base—rate problems, there is also the repeated measurement factor, **Iteration**, as each subject worked on the same problem three times.

4.5.4. DEPENDENT VARIABLES, MEASURES, AND RATING PROCEDURE

For each of the three problems, subjects produced written notes, a reproduction of the story and the question, an answer, and a protocol of considerations and justifications for the answer. Responses on the postexperimental questionnaire on the third trial (homework) including another extensive justification, the test in probability, and some biographical data were also obtained.

Subjects' notes, reproductions of the problem, and reproductions of the text were rated for correctness by the author. These three texts were rated incorrect if they contained definitely false information. The rating procedure for the analyticity versus intuitiveness of the justification was identical with that described in Chapter 3.3.4. Four raters judged each justification blindly (i.e., the judges did not know the independent variables for each subject when rating) and independently according to the first eight dimensions of the above features and attributes of analytic or intuitive thinking (see Table 3.1.). A three level rating was given, with the scores intuitive, analytic, and non-classifiable/ intermediate. In Experiment B, Feature I (feeling of certainty) showed an extremely low interrater reliability. In the present experiment, it was scored directly according to the subjects' confidence rating on a 6–point scale, with the poles "not at all", and "very" (certain). A rating of 1, 2, or 3 yielded an intuitive, and a rating of 4, 5, or 6 an analytic score. It should be noted that the codings were produced by the same raters as in Experiment C. Hence, it is also possible to arrive at some conclusions about the intrarater reliability. The justifications were also examined to ascertain how many words and cues they contained. A cue is defined as a single argument, or as a step within an argumentation or proof that contains an essential transformation. The effects of the independent variables on the response behavior will be investigated when analyzing frequencies of diagnosticity and 'normative responses'. The parametric $|d_2|$ measure (cf. Chapter 2.6.4.) which normalizes the deviations in base–rate problems with different parameters will also be applied.

Although the data were collected according to an analysis of variance design, no (multivariate) parametric tests were calculated. This decision was made because of the low scaling level of most dependent variables and the limited robustness of the analysis of variance against violations of its assumptions when dealing with samples of the size found in the present experiment (cf. WINER, 1971; BORTZ, 1985; DIEHL, 1978). However, the above factorial design is also advantageous for nonparametric (univariate) analysis, as due to the random assignment, one may expect balanced distributions (e.g., of sex) when testing hypotheses (e.g., on the impact of the framing variable, although the effects are then usually weakened by a higher intracell variance). Nevertheless, if the iterated univariate nonparametric tests are applied, the alpha error increases through the repeated testing. If one wants to fulfill all the requirements of classical hypothesis testing, it would be necessary to design an adequate instrument for adjusting the alpha error (cf. Chapter 2.6.5.). As this was not carried out, the results have to be interpreted with some reservations.

4.5.5. RESULTS

Accuracy of text reproduction. The first checks were made on subjects' text processing and text comprehension. Within the total of 32 subjects, there were five notes, four reproductions, and four questions that were rated as incorrect on the first trial on the base−rate problem. One subject did not make any notes on the first trial. On the second trial, there was still one incorrect note, four incorrect reproductions of the story, and one incorrect reproduction of the question. Seven subjects did not make any notes on the second trial. No subject was judged to have produced incorrect reproductions for all three texts (i.e., notes, reproduction, and question).

Table 4.2. presents the first trial frequencies of transformations of the diagnosticity information into an outcome−specific meaning. There are significantly more transformations documented within the notes ($Chi^2 = 16.2$, $df = 1$, $p < .001$) and reproductions ($Chi^2 = 7.4$, $df = 1$, $p < .001$) on the TV problem.

An increasing penetration into the problem structure can be seen with the iterations. Whereas in the first trial only six subjects provided formulations in which an outcome−specific meaning of diagnosticity information was coded by Rater 1, this number increased through 11 up to 19 out of 32 within the third trial. Here COCHRAN's Q indicates significance, $p < .01$. It should be noted that again most of the transformations were observed in protocols on the TV problem, as only one transformation could be identified in the first two trials on the TV problem, and five in the third trial.

When looking at the mistakes in the reproductions, there are two well−known candidates for source of error (cf. Chapter 3.3.5). First, in the TV problem, one subject integrated the diagnosticity into the base−rates and supposed the base−rate to be assessed with the defective test device. Second, the diagnosticity in the Hit problem was interpreted by another subject as the matching or hit probability (i.e., 80% of the predictions actually are true).

Response distributions and length of justifications. Although the sample size is rather small, one may state that the subjects produced a typical response distribution for postgraduate students. If the problems are pooled, there are 41% diagnosticity responses, 12% extreme errors, and 34% middling responses given on the first trial. These values are very similar to those obtained in Experiment A with a comparable group of postgraduate students (cf. Table 2.3., Age 4).

CHAPTER 4.5.5.

Table 4.2.: First trial frequencies of transformations of the diagnosticity information into outcome−specific meanings in notes, reproductions of text, and justifications (number of subjects per cell in brackets)

	Cover Framing	
	Social Judgment (Hit Parade)	Problem Solving (TV)
Notes	2(15)	14(15)
Reproductions	5(15)	14(16)
Justifications	1(16)	5(16)

The attempt to elicit extensive justifications by applying a painstaking and expensive experimental procedure proved to be rather successful. Obviously subjects were willing to cope intensively with the base−rate problem in the second and also in the third trial. The mean number of words or mathematical signs included in the text were; 75.0 in the first, 92.9 in the second and 229.1 in the third trial. This increment is highly significant as a KENDALL's Tau = .69, p < .01, indicates.

Interrater reliability by the classification of the modes of thought. Table 4.3. presents the frequencies of analytic and intuitive ratings on the single features for the first trial. If all raters and features are pooled, there is an average of 43% analytic, 24% intuitive and 33% nonclassifiable intermediate scores. These numbers are almost identical with those obtained with the same sample of raters when coding Experiment B. There seem to be fewer intuitive ratings on the Features E, F, and H in the first trial responses (compared to Experiment B), but for the Majority rating, these differences are not statistically significant. The number of intuitive ratings are higher in Trials 2 and 3, and also on about the same level as in Experiment B if all trials are pooled. As may be inferred from further comparisons of Table 3.3. with Table 4.3., the characteristics of the raters and the features in respect to the frequencies of codings are reproduced in the present experiment.

Before turning to the results on the reliability of strategy classification, we will report the findings on how the confidence feature (Feature I) fits into the intuitiveness −analyticity scale. This feature proved to be extremely unreliable in Experiment B, and showed an extreme outlier characteristic. Consequently

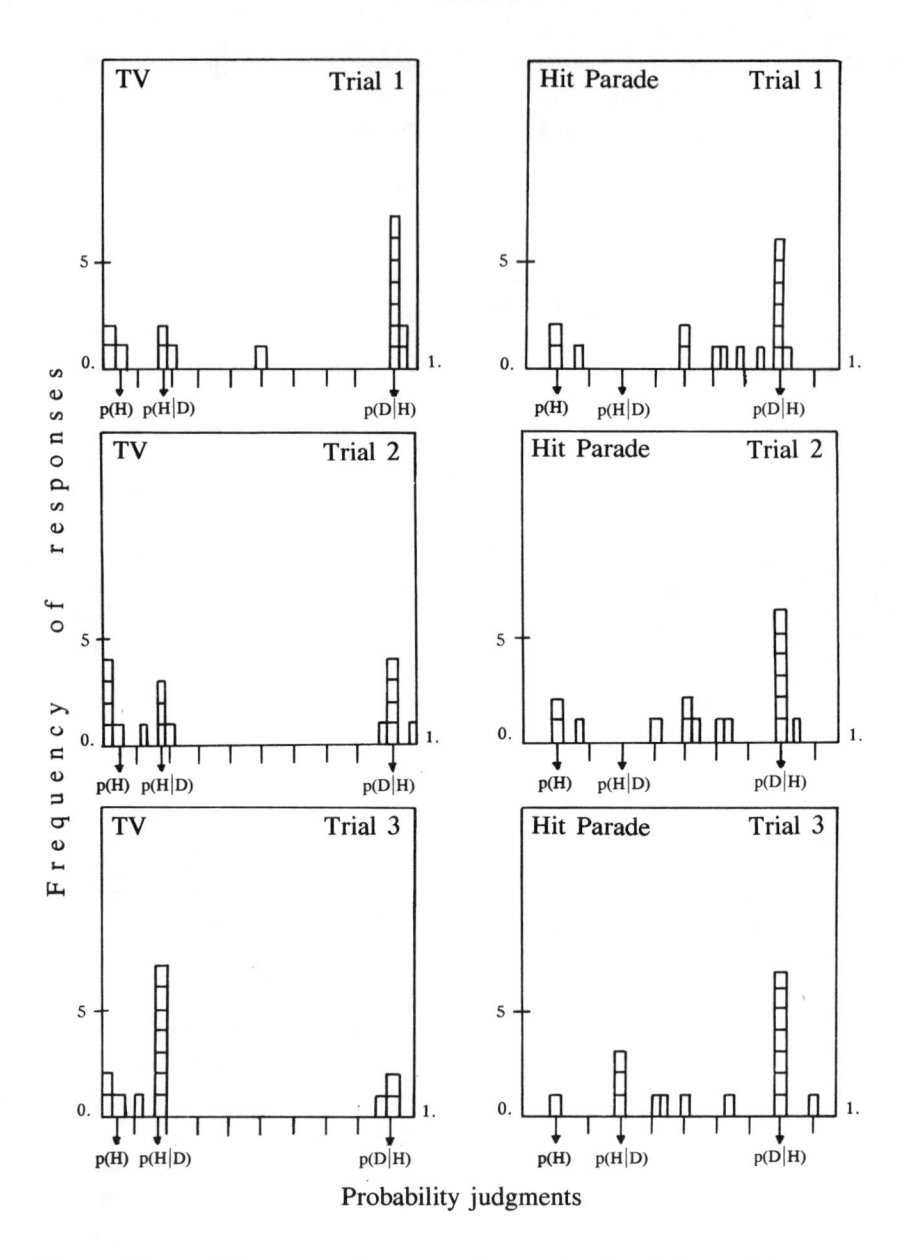

Figure 4.1.: Histograms of response frequencies for all three trials
(left: TV problem; right: Hit Parade problem)

Table 4.3.: Frequencies of analytic and intuitive scores for Features A to H and the strategies according to the four raters' judgments (N=32)

Features	A	B	C	D	E	F	G	H	$\bar{x}$
Rater				Analytic scores					
1	17	19	8	4	24	12	12	19	14.4
2	18	17	4	8	23	6	13	21	13.8
3	15	12	11	3	27	13	14	21	14.5
4	16	13	14	8	22	10	8	9	12.5
				Intuitive scores					
1	13	10	11	8	1	3	14	2	7.8
2	10	10	15	4	2	2	11	8	7.8
3	10	10	10	11	1	2	7	2	6.6
4	12	9	9	17	4	4	14	3	9.0

Table 4.4.: Percentage of matching scores in strategy ratings, a) including nonclassifiable/intermediate ratings, and b) excluding nonclassifiable/intermediate ratings

a) Rater	1	2	3	4		b) Rater	1	2	3	4
1	.	.78	.94	.69		1	.	1.00	1.00	.76
2	.78	.	.81	.63		2	1.00	.	.88	.73
3	.94	.81	.	.66		3	1.00	.88	.	.84
4	.69	.63	.66	.		4	.76	.73	.84	.

we tried to measure it directly with a self-rating procedure. Through all raters, iterations, and applied methods of cluster analysis (cf. Chapter 3.3.4.), we obtained 90 different dendrograms. The analysis of the dendrograms revealed that Feature I showed the same outlier characteristic as in Experiment B. As an illustration, we present the dendrogram for the Majority rating using WARDs' method (cf. Figure 4.2.). We also controlled how the inverted (i.e., negatively poled) confidence scale fitted into the analyticity–intuitiveness scale. However this feature showed the same outlier property as Feature I itself. Because of these results, we cancelled the confidence feature from the analyticity–intuitiveness scale, as already projected in Chapter 3.

Table 4.5.: First trial Z–values for the single features, Z– and Z_{max} values for the strategy ratings for all pairs of raters, and mean Z–values ($\bar{Z}$) for all three trials pooled across all three raters

Features Pairs of raters	A	B	C	D	E	F	G	H	Strat.	Z_{max}
1 x 2	.83	.87	.53	.34	.88	.43	.76	.76	.87	.94
1 x 3	.80	.78	.85	.38	.91	.67	.72	.77	.97	.97
1 x 4	.62	.78	.61	.32	.82	.76	.50	.42	.76	.84
2 x 3	.79	.76	.50	.32	.87	.34	.67	.73	.87	.88
2 x 4	.68	.76	.71	.48	.82	.36	.57	.59	.73	.81
3 x 4	.68	.72	.68	.35	.81	.62	.42	.57	.73	.81
$\bar{Z}$-1st. trial	.73	.78	.65	.37	.85	.53	.61	.62	.82	
$\bar{Z}$-2nd. trial	.59	.58	.60	.37	.77	.50	.54	.51	.74	
$\bar{Z}$-3rd. trial	.69	.79	.67	.47	.83	.42	.69	.59	.77	

In the following analysis of the interrater reliability, we will discuss the first trial interrater reliability in detail, and only report some of the results of the second and third trials, placing particular emphasis on deviations from the results of the first trial. The first trial's matching scores of strategy ratings are reported in Table 4.4. There is an average of 75% matchings if the nonclassifiable strategies are included, and of 87% if only those subjects are considered who have been classified by both raters in each pair. These values are an improvement on those obtained in Experiment B. (For the following cf. Table 4.5.) The first trial average Z-value of strategy classification is also above the .8 critical value. The second and third trial values of matching scores and the Z-value are slightly lower, and about the same as the values found in Experiment B. Due to the relatively small sample size of 32 subjects, however, small variations are to be expected. For instance, Rater 4's first trial mean Z-value was with .68 the lowest, as in Experiment B. However in the third trial he scored much better, resulting in an average Z-value of .81, whereas Rater 3's average Z-values decreased from .86 in the first, through .74 in the second, ending up with .67 in the third trial.

The single feature average Z-values are mostly lower than the Z-values obtained from the strategy ratings. Only Feature E's (dependent on vs. independent

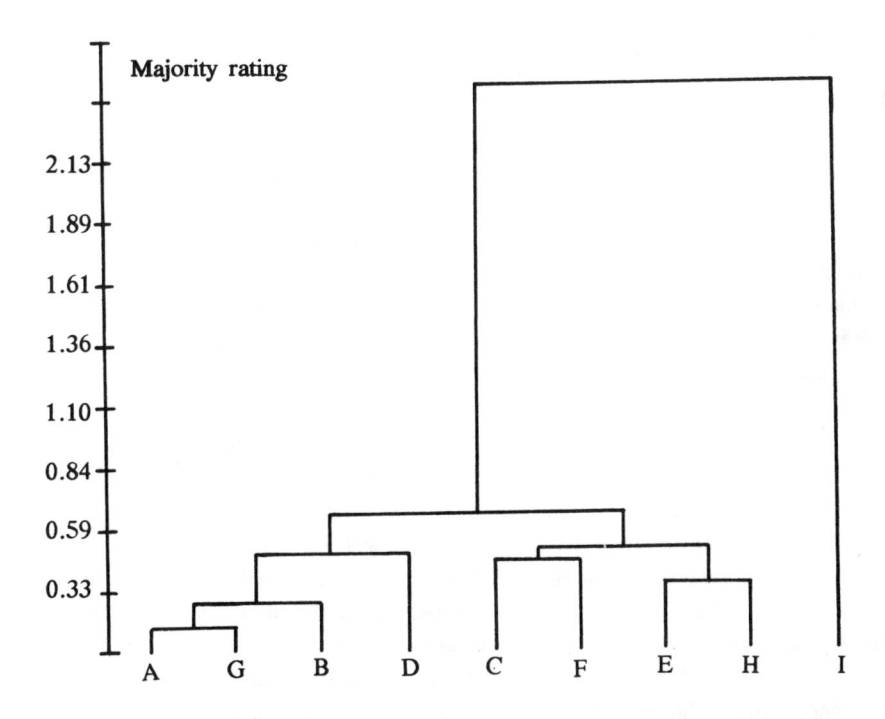

Figure 4.2.: Dendrogram of the intuitiveness-analyticity features (Group average method) for the Majority ratings on the first trial

of personal experience) Z−value is higher, which is presumably caused by the high percentage of diagnosticity ratings. As in Experiment B, the average interrater reliability for Feature D (treating the problem structure as a whole vs. separating details) is rather unsatisfactory and only marginally above chance level in the first and the second trial. The average on the third trial is somewhat better. If Rater 3 is excluded for instance, an average of .68 is obtained which approximates the mean Z−value for the single features. Attempts to increase the interrater reliability by eliminating the features with the lowest Z−values do not result in a higher average of matching scores, hence the Feature D and also the somewhat critical Feature F will be retained in the following hypothesis testing procedures.

We will not give a detailed report on the results of the exploratory cluster analysis on the structure of the intuitiveness − analyticity scale, but only mention that the patterns revealed in Experiment B could be replicated. As the structure of the clustering remains unaffected by the inclusion of the Feature I (due to its outlier characteristic), some evidence for this finding is exemplarily provided by Figure 4.2. Once more, a cluster of features on the cognitive aspects (A, B, G) may be identified which is tied to the Feature D and which is somewhat separated from the features that reflect personal and emotional aspects. It should be noted that through the three trials this separation disolves in favor of a chaining of all features.

The effect of framing on the mode of thought and strategies. The effect of the Experimental Framing and of the Cover Framing on the mode of thought was tested with a MANN−WHITNEY U−test on the mean sum scores of intuitive−analytic features across all raters and subjects. Table 4.6. presents the means for all four framing conditions. For both framing variants and all three trials, more analytic scores are attributed in a Problem Solving Frame than in a Social Judgment Frame, as proposed in *Hypothesis 1*. The inverse statement holds true for the intuitive scores which correlate negatively with the analytic scores. However, the inferential statistic only yields a marginally significant difference for the analytic scores in the second trial, $p < .07$, and a significant difference, $p < .05$, for the third trial's analyticity ratings, for both the comparisons in the **Cover Framings**. The effect of the Cover Framing seems to increase with the iterated working on the problem. However, FRIEDMAN's analysis of variance did not produce significant results. According to the Majority rating, the same consistent pattern is produced by the frequencies of analytic and intuitive strategies. For instance, once more, there are more analytic strategies under a Problem Solving Cover (the TV problem) than under a Social Judgment Cover (the Hit Parade problem) in all three trials. The same anticipated relation may be observed for the **Experimental Framing** variable, though the effects

Table 4.6.: Mean scores of analytic and intuitive features in the different framing conditions averaged across the four raters (Social Judgment abbreviated to Soc., Problem Solving to Prob; the label '(both)' means that the framing conditions are pooled)

Exp. Fram.	Soc.	Soc.	Prob.	Prob.	Soc.	Prob.	(both)	(both)
Cover Fram.	Soc.	Prob.	Soc.	Prob.	(both)	(both)	Soc.	Prob.
First trial								
Analytic	3.2	3.2	3.9	4.1	3.2	4.0	3.6	3.7
Intuitive	2.5	1.8	1.7	1.4	2.2	1.5	2.1	1.6
Second trial								
Analytic	2.1	3.8	3.6*	4.0	2.9	3.8	2.8	3.9
Intuitive	3.1	1.3	2.0*	1.3	2.2	1.6	2.5	1.3
Third trial								
Analytic	1.7	4.6*	3.9	4.5	3.2	4.2	2.8	4.6
Intuitive	3.5	1.4*	2.0	1.5	2.5	1.7	2.8	1.4

*) Due to one missing protocol the cell frequency is only 7.

are slightly below statistical significance. The effect of Experimental Framing is strongest in the first trial, but is only marginally significant, $p < .10$. As may already be conjectured by an analysis of Table 4.6., there are no statistical interactions of the two framing variants. Yet, if the 'extreme group' of a 'double' Problem Solving Frame is compared to the 'double' Social Judgment Frame, the effects of framing variants add up, and the U−test indicates significant results, $p < .05$, for the second and third trials on the analyticity scores. One should note that the two framing variants are independently varied, and hence the 'double' framing may be regarded as a higher degree of framing than a mixed framing.

Summing up all the results on the applied U−tests while taking the consistency and the relative small power of the applied tests into consideration (the later is to some extent due to the relatively small sample size; for an analysis of the impacts of consistency and power see e.g., STELZL, 1982, or WOLINS, 1982), *Hypothesis 1* can be considered confirmed.

The expectations regarding the **frequency of extreme answers** may also be regarded as confirmed for the **Experimental** Framing variable (cf. Table 4.7.). There are significantly more extreme errors in the first trial in an Experimental Problem Solving Frame than in an Experimental Social Judgment Frame, $p < .05$, FISHER−

Table 4.7.: Parameters of response distributions in the different trials split according to Experimental Framing (absolute frequencies in brackets)

Experimental Framing	Problem Solving	Social Judgment
	$|d_2|$ error measure	
Trial 1	26.1	19.0
Trial 2	25.2	20.5
Trial 3	14.2	19.8
	Normative solutions in %	
Trial 1	6(1)	6(1)
Trial 2	13(2)	6(1)
Trial 3	38(6)	25(4)
	Diagnosticity responses in %	
Trial 1	38(6)	44(7)
Trial 2	19(3)	38(7)
Trial 3	25(4)	31(5)
	Extreme responses in %	
Trial 1	38(6)	6(1)
Trial 2	31(5)	13(2)
Trial 3	13(2)	19(3)

test. Due to the slightly more extreme base–rate and diagnosticity parameters, the area of extreme answers in the Problem Solving **Cover** Frame (TV problem) is only 57% of that in a Social Judgment Cover Frame (Hit problem). Nevertheless, there are (nonsignificantly) more extreme answers in the Problem Solving condition (cf. Table 4.8.)!

The percentage of diagnosticity responses did not significantly differ between the various framing conditions (cf. Table 4.8. and 4.9.). The percentage of 'normative solutions' and the $|d_2|$ measure did not differ significantly between the framings either (i.e., Problem Solving Cover vs. Social Judgment Cover and Experimental Social Judgment vs. Experimental Problem Solving Frame).

We close the section on the influence of framing effects with a report on the results of the postexperimental questionnaire. The subsequent levels of significance are determined by the nonparametric MANN–WHITNEY U–test, and throughout the next two paragraphs the medians of the response distributions will be reported. In the **Experimental** Problem Solving Frame, subjects spent more time on their

Table 4.8.: Parameters of response distributions in the different trials split by Cover Framing; absolute frequencies in brackets

Cover Framing	Problem Solving	Social Judgment
	Probabilities	
Base Rate	.2	.10
Diagnosticity	.90	.80
Normative solution	.155	.31
	$|d_2|$ error measure	
Trial 1	23	22
Trial 2	26	20
Trial 3	16	18
	Normative solutions in %	
Trial 1	12(2)	0(0)
Trial 2	19(3)	0(0)
Trial 3	44(7)	19(3)
	Diagnosticity responses in %	
Trial 1	50(7)	38(6)
Trial 2	24(4)	38(6)
Trial 3	12(2)	43(7)
	Extreme responses in %	
Trial 1	25(4)	19(3)
Trial 2	31(5)	12(2)
Trial 3	19(3)	13(2)

homework than subjects in the Experimental Social Judgment Frame, 61.2 minutes compared to 45 minutes, $p<.05$; worked about twice as long with full concentration on the homework, 39.2 vs. 20.8 minutes, $p<.05$; judged the content of the stories to be less interesting, 3.7 vs. 4.8, $p<.07$; judged themselves to have been more intensely involved in their work, 4.9 vs. 4.5, $p<.09$; and judged the written justifications to be significantly less able to convey their thoughts, 4.3 vs. 5.0, $p<.02$, than subjects in the Experimental Social Judgment Framing.

The effect of the **Cover** Framing was less extreme. Although the total time of homework was about the same under both framing conditions, 58.3 compared to 58.1 minutes, the time of concentrated working was marginally significantly higher in the Problem Solving than in the Social Judgment Frame, 30. vs. 21.2, $p<.09$. It should also be noted, that in the Problem Solving Cover, ten subjects, as opposed to six in the Social Judgment Cover, talked with a third person about their homework, and that marginally significantly more subjects in the former judged their responses to be influenced by a third person, 7 vs. 2, Chi^2=3.05, df=2, N=28, p<.10.

Iteration. In the course of repeated working on the base−rate problem, there was no significant shift toward analytic or intuitive features. Across all subjects and across all raters, the means of analytic features were 3.6 in the first, 3.4 in the second, and 3.7 in the third Trial. The mean scores of intuitive features were 1.9 in the first two trials and 2.1 in the third Trial. Likewise, no interaction between the iteration variable and the framing variants or the other independent variables could be found with respect to the features of analyticity or intuitiveness. Thus with respect to the shift toward analyticity, *Hypothesis 2* cannot be confirmed. Although there is no shift in the intuitive−analytic rating, the number of words (see above) and cues contained in the justifications increased in an almost deterministic manner through the iterations. For instance, 22 subjects reported more cues in the second trial than in the first, but only one subject less. In our sample, there were 27 justifications which contained more cues in the homework, but none containing less than the number present in the second trial. If all subjects are pooled, a slight but significant increment of confidence ratings is also recorded. Through Trials 1 to 3, the medians were 4.0, 4.0, and 5.0, resulting in KENDALL's Tau=.76, p<.01.

There seems to be some improvement in the behavioral data through Trials 1 to 3. If one only studies the percentages (cf. Table 4.7 or Figure 4.1), there was a decrease in diagnosticity responses, and an increase in middling responses. The latter is clearly due to the increment of 'normative' answers. However, presumably because of the low sample size, nonparametric analyses of variance for the iteration variable on the various d or |d| measures (cf. 2.6.4.) did not indicate a significant trend in the behavioral data. The increase of 'normative' answers in the third trial compared to the first or second were the only significant effects in pairwise comparisons (!). A McNEMAR−test provides p<.01 for Trial 1 vs. 2, and p<.001 for Trial 1 vs. 3.

Career Socialization. For the factor Career Socialization no significant effects could be found in either the postexperimental questionnaire or the biographical data (such as age, type of prior school education, etc.). The average student in Natural Science usually receives a more thorough mathematical and stochastic instruction than the average student of Social Sciences. A control of this was given by the screening test in stochastic methods. The Natural Scientists did significantly better on all six items in this test, and consequently rated their ability to treat the problems in the experiment, which to some extent are affiliated to the test items, higher than the Social Scientists. A median of 4.5 was found in the Natural Science sample compared to 2.9 for the Social Scientists.

Table 4.9.: Parameters of response distributions split by Career Socialization (absolute frequencies in brackets)

Career Socialization	Social Science	Natural Science
	$|d_2|$ error measure	
Trial 1	22.8	22.2
Trial 2	24.9	20.8
Trial 3	20.4	13.5
	Normative solutions in %	
Trial 1	0(0)	12(2)
Trial 2	0(0)	19(3)
Trial 3	19(3)	44(7)
	Diagnosticity responses in %	
Trial 1	31(5)	50(8)
Trial 2	25(4)	38(6)
Trial 3	25(4)	31(5)

More Natural Scientists responded with the 'normative solution', and in their homework, 7 out of 16 subjects in this group actually produced such an answer (cf. Table 4.9.). But the percentage of diagnosticity responses was also higher in the Natural Scientist sample! Although decreasing from 8 out of 16 to 5 out of 16, it was still higher than the number of diagnosticity responses in the Social Science group who produced 5 out of 16 in the first, and 4 out of 16 in both the second and third Trials. Only one Natural Scientist but five Social Scientists supplied a middling response that was not the 'normative solution' in their homework. Before turning to the modes of thought, we will report the results on the $|d_2|$ measure. The performance of the Social Scientists did not improve through the trials, whereas there was definite improvement for the Natural Scientists (KENDALL's tau = .86, $p < .01$).

Although there were no very strong effects in the behavioral data, the **modes of thought** were strongly **affected** by the factor Career Socialization. The frequencies of the applied mode (and also the frequencies of algebraic strategies) strongly covaried with Career Socialization, according to the Majority rating in the first two trials. Natural Scientists showed almost exclusively analytic (and mostly algebraic solutions) whereas for the Social Scientists, there were about an equal number of intuitive and analytic strategies (and a majority of nonalgebraic solutions, cf. Table 4.10) through all trials. Through all three trials, there were significantly more analytic features and significantly fewer intuitive features assigned to the Natural Scientists compared to the Social Scientists, as stated in *Hypothesis 3*. The values for the differences in the frequencies of algebraic strategies between the two career socializations through Trials 1 to 3 are: $Chi^2 = 3.33$, $Chi^2 = 1.63$, $Chi^2 = 4.66$, $df = 2$ for all three tests, indicating marginal significance in the first, and significance in the third trial. The typical response strategy of a Natural Scientist is analytic — algebraic. According to the Majority rating, in the first trial, half of the subjects in this sample chose such a strategy, compared to only 12% of the Social Scientists. However, about 45% of the Social Science students were rated as having chosen an intuitive — nonalgebraic approach, which was only found by two out of the sixteen Natural Scientists. Finally, the Natural Scientists showed greater confidence in their homework ($p < .07$, U — test).

Table 4.10.: Mean scores of analytic and intuitive features across all four raters (a), frequency of analytic and intuitive strategies (b), and frequencies of algebraic and nonalgebraic strategies (c), split by Career Socialization; Social (Soc.) vs. Natural (Nat.) Science students

(a)	Analytic scores		Intuitive scores	
	Soc.	Nat.	Soc.	Nat.
Trial 1	2.2	4.9	2.7	1.0
Trial 2	2.4	4.4	2.7	1.1
Trial 3	2.4	5.0	3.0	1.2

(b)	Analytic strategies		Intuitive strategies	
	Soc.	Nat.	Soc.	Nat.
Trial 1	7	13	7	2
Trial 2	8	13	7	1
Trial 3	5	11	7	4

(c)	Algebraic strategies		Nonalgebraic strat.	
	Soc.	Nat.	Soc.	Nat.
Trial 1	3	9	13	7
Trial 2	4	8	12	8
Trial 3	6	13	10	3

Sex: No significant effect of Sex could be found in the inferential statistics on the behavioral data. The pattern of frequencies not only showed no significant differences but was extraordinarily similar. However, the sum scores of the analyticity — intuitiveness rating were affected by the Sex variable (cf. Table 4.11.). Contrary to our expectations, the women's justifications were rated to be more analytic than the men's. In the first trial, the difference between the mean analyticity ratings attained statistical significance. The intuitiveness scores for the first and second Trials showed a marginal significance between the sexes.

Table 4.11.: Mean scores of analytic and intuitive features across all four raters split by Sex

	Analytic score Male	Female	Intuitive score Male	Female
Trial 1	2.7	4.4	2.4	1.3
Trial 2	2.7	3.9	2.5	1.4
Trial 3	2.7	4.0	2.4	1.9

Effect of mode of thought on response distributions. The modes of thought generate different response distributions. Information on the response distributions for all three repetitions are presented in Figure 4.3. In this figure the d_2 measure is used as the different problems are pooled. When the first trial is considered, it can be seen that half of the responses in the analytic mode were diagnosticity answers, whereas only three diagnosticity responses (33%) occured in the intuitive mode. If the absolute numbers are considered, in the first trial there were significantly more diagnosticity responses produced by the analytic mode than by the intuitive mode ($p<.05$, Binominal−test). If the relative frequencies are regarded, the difference is below statistical significance and vanishes through the iteration. Many subjects were capable of determining the 'normative solution' in their homework when using the analytic mode, and only 2 out of a total of 16 responses retained the diagnosticity in the homework. The predominant response type in the intuitive mode was the middling response. In the first trial two−thirds of the intuitive responses were of this type, but only a fifth of the analytic answers, which fraction included two 'normative solutions'. No 'normative' answers could be observed in the intuitive mode, whereas the percentage of 'normative' answers in the analytic mode increased from 10%, through 19%, to 62%.

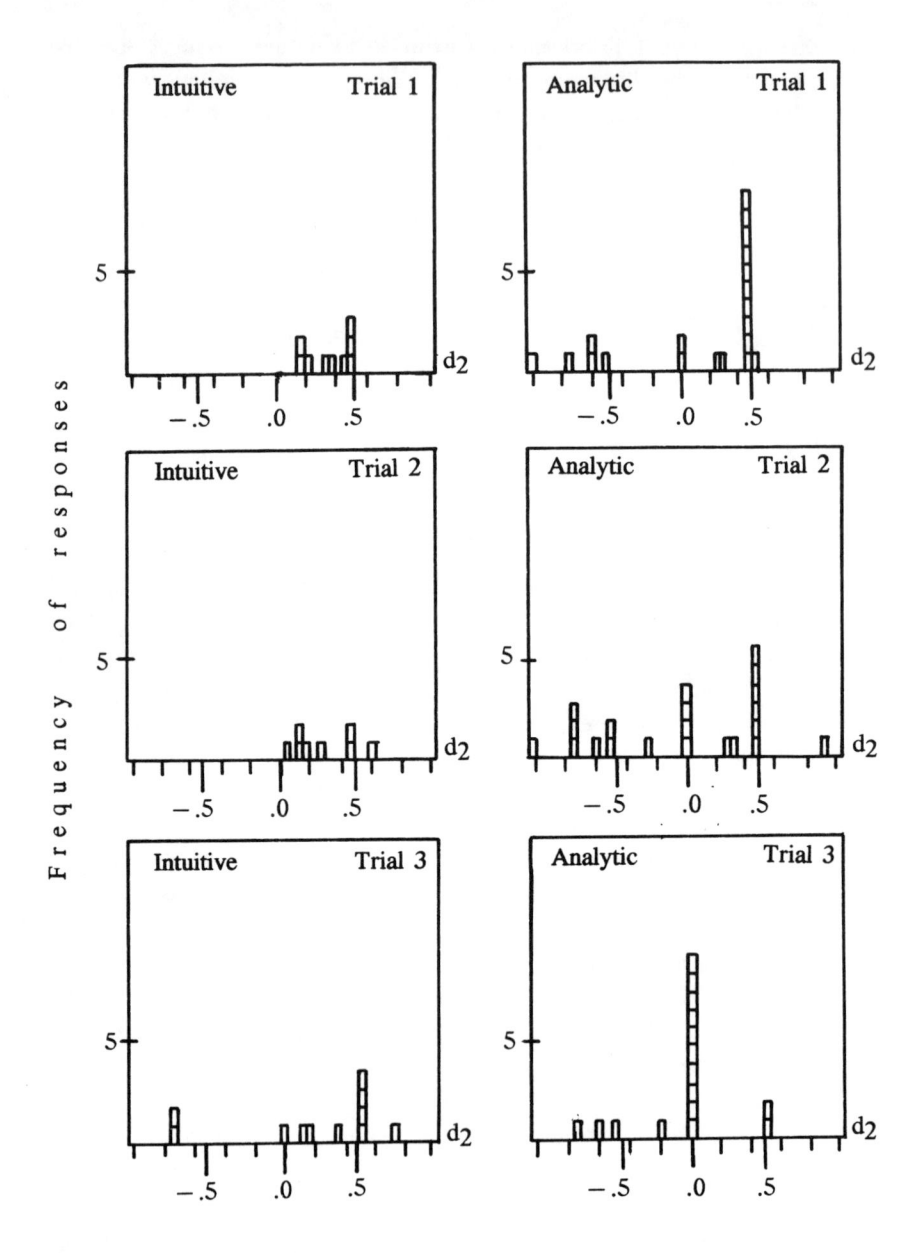

Figure 4.3.: Values of d_2 split for the intuitive and analytic mode of thought according to the Majority rating for all three trials

4.6. DISCUSSION OF EXPERIMENT C'S RESULTS

Literal text acquisition does not imply any cognitive unfolding of a problem structure. In individual sessions, subjects were repeatedly obliged to listen to, to read, and to produce complete rewritings of a whole base−rate story and question. Compared with Experiment B, a far higher percentage of subjects produced **correct reproductions**, although the text was no longer visible during the reproduction phase and had to be recalled with own notes as the only reference. However, a correct literal reproduction by the subjects does not imply an analytic penetration into the problem structure.

Subjects' difficulties in appropriately using the base−rate information in the Hit problem have been observed before (cf. Experiments A and B). Theoretically this might be because in the base−rate problems that are presented in these experiments (and likewise in many others), the outcome−specific meaning of the diagnosticity had to be inferred from the text. Because of this, the explicit meaning of the diagnosticity information was included in the problem presentation for Experiment C (see Chapter 4.5.3.).

Despite this, less than one third of the subjects explicitly actualized an outcome−specific decoding of the diagnosticity information in the Hit problem. An outcome−specific comprehension of the diagnosticity information (i.e., 80% correct predictions, both for the songs that became a hit and for those that did not, means 80% hit predictions in the case of a hit, and 80% non−hit predictions in the case of a non−hit), seems to be a crucial barrier in base−rate problems which sometimes seems to be hard to overcome (for instance in the Social Judgment Cover Frame). Consequently, tutorials and instructions on coping with base−rate problems (cf. BEYTH−MAROM & LICHTENSTEIN, in press; LICHTENSTEIN & McGREGOR, 1984) will have to focus on this step.

In contrast to the Social Judgment Cover (i.e., the Hit Parade problem), subjects were obviously able to cope fairly well with the Problem Cover (i.e., the TV problem) in Experiment C. Not only was the transformation of the diagnosticity information included in most of the notes, reproductions, and justifications, but also a considerable proportion of the subjects produced a 'normative' response. The problem structure of the TV problem seems to be somewhat more accessible; this is reflected by a longer time of dealing with the problem, and a more highly rated intensity of working compared to the Social Judgment Cover. But obviously, the subjects preferred to deal with the Social Judgment Cover.

The question as to how these behavioral differences and the different percentages of 'normative solutions' between the two stories may be explained other than by the (content) framing features, can possibly be clarified by the reversed time order in the Hit problem (outcome subsequent to diagnosticity). This might be regarded as another fundamental source of difficulty.

The lists of features provide a reliable procedure for classifying strategies. The lists of features which were used for the rating procedure on strategy classification were developed when facing Experiment B's data. Hence the interrater reliability achieved in that study might have been caused by the author's knowledge of the data structure and the specific contents of the written protocols. However, Experiment C provides a critical test on the reliability. The interrater reliability obtained in this study was on the same high level and even slightly better in the first trial, for instance, resulting in a Z−value of .82.

The substructure of the analyticity−intuitiveness scale was also replicated in this experiment (cf. Chapter 3.), and the features which point at the inferential cognitive side have been clustered. Another cluster containing the features which point at the emotional side may also be identified. The confidence feature revealed a rather outlier characteristic in Experiment B, and has further proved to be very unreliable. We attempted to measure the confidence directly by letting the subjects rate it themselves. The multivariate cluster analysis revealed the same outlier characteristic for this feature by each of the four raters' and also the Majority's coding. This outlier characteristic was also maintained for the inverted confidence scale, that means, if confidence in the answer is assumed to be tied to analyticity instead of intuitiveness. As the confidence scale showed an idiosyncratic characteristic, we cancelled this feature from the lists. Using a model of the process and the structure of the cognitive activity in stochastic thinking as a base, we will discuss different aspects of confidence in Part II of this text (cf. Chapter 5.), and will also provide a framework for the process and structure of human information processing. This framework allows for an understanding of the interrelationship of the single features but also provides a theoretical basis for the derivation of hypotheses on the mode−specific characteristics of the features.

Framing affects mode of thought and response distributions. If all responses are pooled, both the **first trial** response distribution and the first trial frequency of the modes of thought are rather similar to the patterns obtained from comparable subject samples on base−rate problems.

The high percentage of diagnosticity responses (41%) in the first trial, and also the frequencies of the other response ranges (except perhaps the 'normative solution', which is not always found in other experiments) are highly comparable (cf. BAR− HILLEL, 1980; or the results obtained in Experiment A). If all trials are pooled, the percentage of strategies judged by the Majority rating to be intuitive (26%) and analytic (62%) (cf. Table 3.8, and Figure 4.2) is also similar to the percentages obtained in Experiment B that are 34% intuitive and 62% analytic (although it must be noted that different age groups were pooled in this study).

According to our theoretical considerations, a Problem Solving Frame more frequently elicits the analytic mode of thought, and a Social Judgment Frame more frequently the intuitive mode. These results may be regarded as a kind of construct validation, and they demonstrate the dependency of the cognitive activity on contextual and framing features.

There is also some evidence that the effect of the Cover Framing increased through the trials. This increased impact is also plausible, as the time subjects spent on thinking about the problem, and thus presumably the extent of the retrieval and the mode–specific construction of a context, was increased by the iterated treatment of the problems.

Prototypical strategies are to be found in a 'pure' framing. If the Cover (or Content) and the (Procedural–) Experimental Framing are both placed within either the Social Judgment or the Problem Solving Frame in the iterations, subjects nearly all shift to the framing–specific modes of thought. With a Social Judgment Cover x Experimental Framing in Trials 2 and 3, most of the strategies were of the non-algebraic–intuitive type, whereas the algebraic–analytic strategy was predominant in the double Problem Solving Frame. These results have a limited validity due to the small sample size. We will present prototypical strategies and deal with the behavioral consequences of the modes of thought below, after first discussing the analysis of the differential variables.

Career Socialization is crucial, but what about Sex? In our study **Career Socialization** is a very crucial variable. Before going any further, one should recall that significant differences could not be detected in the postexperimental questionnaire on either the perception of or the dealing with the experimental tasks, or in age or number of years spent at school. Nevertheless, many characteristics of the response distributions and of the modes of thought do strongly differ in the following ways.

The response pattern of the Natural Scientists is more restricted than that of the Social Scientists. In all three trials, more than half of the Natural Scientists responded with the diagnosticity or the 'normative solution', ending up in the homework with 75% producing one of these two responses. Consequently, only singular middling responses could be found in the Natural Scientists' sample. In the following remarks, one should recall that in all three trials, the Natural Scientists produced more diagnosticity responses, and that in the first trial the average normalized deviation ($|d_2|$; see Table 4.9.) from the 'normative' answer was only slightly lower than that of their Social Science colleagues. But obviously, many of the Natural Scientists were able to find the 'normative solution' in the course of iterated dealing with the base—rate problem, as 44% provided such a response in their third trial. Thus, the average deviation of the Natural Scientists became much smaller through the trials, whereas the average deviation in the Social Scientist group did not decrease.

The ways in which the probability judgments were assessed showed a more differential character. Natural Scientists commonly proceeded in the (algebraic—) analytic mode. Of course, this mode is almost of necessity the most frequent mode of thought in the third trial, if one takes into consideration that this trial yielded a high number of 'normative solutions'. But even in the first trial, the Natural Scientists were mostly judged to be operating in this mode, although this trial produced very few 'normative' responses (but many extreme errors). On the other hand, (algebraic—)analytic strategies were seldom observed in the Social Scientist sample, in which the (nonalgebraic—)intuitive strategy type was clearly the most frequent one. Of course, the above findings do not mean that this effect is solely caused by the actual career socialization, as due to a very stringent a priori matching, the factor Career Socialization might be confounded with other variables.

However, when interpreting the differences observed in career socialization, we consider the literature on pedagogical codes and the sociology of knowledge (cf. PFEIFFER, 1981; BERNSTEIN, 1977; YOUNG & WHITTY, 1977) to be helpful. Students become socialized at school, socially accepted and permitted behavior is taught, and in higher education, the way of organizing and communicating knowledge is adjusted in respect to their peers and to academic norms. Mathematics and analytic thinking are of central importance in Natural Scientists' socialization (cf. PFEIFFER, 1981; REISS, 1979; STEINER, 1984). Quite obviously, a majority of the Natural Science students (at Bielefeld University, who are more competent at, and more subjectively oriented toward mathematics; this has actually been controlled by WELZEL, 1984) looked for a precise, unique, **algebraic** solution, either in the 'normative solution' or the diagnosticity, and avoided fuzzy middling responses or rough probability judgments.

Like Career Socialization, the factor Sex did not reveal statistically significant effects by the biographical information and the postexperimental questionnaire. Also, no statistically significant effects could be detected in our data on the behavioral distributions in respect to Sex. However, the activated modes of thought and the mean number of analytic/intuitive features differed between the two sexes. When planning the study, we did not expect Sex differences in either the behavioral or the modes of thought measures if the age and educational level were held constant. We were thus somewhat surprised, when in contrast to BASTICK, 1982, and others, the Majority rating revealed more analytic features in the women's strategies than in the men's. We cannot offer a straightforward explanation for this effect. However, this finding may be considered as counterevidence to the widespread belief in a stronger male proneness toward analyticity if the level and faculty of education are matched.

Getting involved in the problem (iterations) yields (some) 'normative solutions', but the diagnosticity responses are maintained. Iterated performance on base–rate problems has been studied by analyzing responses using the same cover story with different parameters (FISCHHOFF, SLOVIC, & LICHTENSTEIN, 1979; LYON & SLOVIC, 1976), or by investigating the responses on packages of various base–rate problems (cf. Experiment A; WELL, POLLATSEK, & KONOLD, 1982; BAR–HILLEL, 1980; HOLT, 1984). However, in these recent studies the text understanding and information processing were not controlled, only a short time was allowed for each probability judgment, and no feedback was given on the responses. Thus, on a behavioral level, no or only slight changes were observed in the response distributions, and hardly any learning effects could be reported.

In the present experiment, subjects were asked to deal with one and the same problem three times. The experimental procedure was arranged in such a way as to encourage considered, sophisticated responses. The way in which the first iteration was introduced, is in some ways similar to FISCHHOFF and BAR–HILLEL's, 1984b, decentralization technique, though our method seems to be considerably weaker. The third trial differed from the second and from other studies. Subjects were sent home and asked to give a well–considered response to the questions.

When tracing subjects' notes, reproductions, and justifications of strategies, the most important finding seems to be the increasing number of incorporations of the outcome–specific meaning of the diagnosticity. The outcome–specific transformation of the diagnosticity has to be regarded as a necessary step in an assessment of the exact 'normative solution'. Thus, the increase in 'normative solutions', the slight decrease in the average (parametric) $|d_2|$ measure, and presumably also the decrease in the diagnosticity responses between the first and the second trials, may be due to this effect. However, as mentioned above, subjects more frequently transformed the diagnosticity, and ended up with the 'normative solution' in the Problem Solving Cover Frame (TV problem) than in the Social Judgment Cover Frame (Hit problem). Due to the small sample size, the interaction of the Iteration and Problem Framing variables was not statistically analyzed.

Summarizing the effects of the Iteration variable on the behavioral data, one may state that only a few postgraduate students produced a 'normative' response on the first trial (although intensively working on the problem), but a considerable proportion will produce such a response, if enough time, etc., is given. This shows that the 'normative solution' is within the scope of students' performance in iterated intensive working on base−rate problems. Yet, if the diagnosticity has already been given as a solution in the second response, it will presumably be maintained on the third trial.

Diagnosticity is a typical (first trial) analytic response. A very surprising finding is that the justifications given with the first trial diagnosticity responses were predominantly rated to be of an analytic type. Thus, the findings of Experiment B could be replicated. The present study also produced little evidence to support the heuristics and cognitive constructs (such as the representativeness heuristic or causal scheme) which are considered to be responsible for a large proportion of the complete neglect of base−rates in base−rate problems (cf. BAR−HILLEL, 1980; TVERSKY & KAHNEMAN, 1979). Most of the justifications given in support of diagnosticity were judged to be of an analytic type. But let us look at some more data. In this chapter we will present elaborate justifications which in our opinion even provide a much better understanding of the subject's proceeding than the protocols in Chapter 2. Due to the length of the protocols, they will only be presented in English translation. The following first trial justification for the TV problem cover was produced by a 24−year−old female Natural Science student participating under a Social Judgment Experimental Frame. She was well−educated in statistics and provided excellent answers in the screening tests. Text and questions were correctly reproduced, including the outcome−specific meaning of the diagnosticity. She argued:

red lamp: lights up for 10% of all faultless tubes
and for 90% of all defective tubes
hence, 9.8% = red lamp and okay
1.8% = red lamp and defect
9.8 : 11.6 = x : 100
980 : 11.6 = 90

The argumentation is close to 'normative', but obviously the subject just fails to determine the correct numerator. According to the above rating procedure, seven analytic and two intuitive scores were assigned on the first trial by Rater 1. The subject was judged to process consciously (A), purely intellectually (B), sequentially ordered (C), independent of personal experience (E), using numerical patterns (F), with high cognitive control (G), and in a cold emotion−free presentation (H). Intuitive ratings were assigned for Feature D, as the problem was treated as a whole, and for the confidence feature (I), due to a high subjective rating. The second trial justification of this subject looks similar, yet resulted in a base−rate response, and in the third trial she finally determined the 'normative solution' using the graphical representation (cf. SCHOLZ, 1981, p. 21) of base−rate problems of the Cab

CHAPTER 4.6.

problem type.

The next example of a diagnosticity response is the text of a 26−year−old male Social Science student produced under an Experimental Social Judgment Frame:

> I have thought out that if 2% of all tubes actually are ? defective and I first of all fix this as 100%; from this 100%, in 10% of all cases the red lamp lights up incorrectly, as it does not work precisely enough. Hence, there remain 90% of all cases, in which the lamp correctly detects a tube−defect.

Here again, the following analytic features were attributed; conscious information processing (A), pure intellectual activity (B), sequential proceeding (C), free of personal experience (E), and cold and emotion−free (H). The high subjective confidence resulted in an intuitive rating, and the other features were not rated. This subject maintained the diagnosticity in Trials 2 and 3.

There is a wide range of strategies. A multitude of strategies are applied in probability judgments on base−rate problems. Both this study and the preliminary Experiment B have clearly demonstrated the existence of two modes of thought. We will introduce two extreme examples. A female mathematics student in the Social Judgment Experimental Frame provided in her first and second Trials a simple, rough, but brief and brilliant analytic justification. Her short text reads as follows:

> I have thought out that in 100 tubes there are 10 faultless tubes with a red lamp and 2 defective ones. Hence, in 12 events, the lamp will light up in only two events in which the tube is really defective, hence the probability is about 20% (The fact that the green lamp lights up in 10% of all cases in which a defective tube will be drawn, is so small that it need may not be considered here).

Analyticity may lead astray. Obviously, the former subject did not rely too much on her argumentation (a confidence rating of only 2 was given on the 6−point scale). She went home, consulted another mathematics colleague, and together they applied the knowledge that they had gained from mathematical statistics. The results of this cooperation read as follows:

> One has to discriminate between four events
> 1. A   red lamp | defective tube
> 2. $\bar{A}$ red lamp | faultless tube
> 3. B   green lamp | defective tube
> 4. $\bar{B}$ green lamp | faultless tube
>
> For our problem only the event, "red lamp lights up, hence the tube is defect" is of interest.
> Let X denote red lamp lights up, Y tube defect. Thus, the probability
> p(tube actually defect: the red lamp lights up) = $p(Y|X)$

$$= \frac{p(X \cap Y)}{p(X)} = \frac{p(X|Y) \cdot p(Y)}{p(X)}$$

$$= \frac{p(\text{red lamp lights}|\text{given tube defect}) \cdot p(Y)}{p(X)}$$

The red lamp lights with a probability of

$$\underbrace{\frac{2}{100}}_{1} + \underbrace{\frac{98 \cdot 10}{1000}}_{2} - \underbrace{\frac{2}{100} \cdot \frac{10}{100}}_{3} = \frac{116}{1000} = p(X)$$

$p(Y)$ = probability that a tube is defect = $\frac{2}{100}$

$p(X|Y)$ = probability that the red lamp lights if the tube is defect = $\frac{2}{100} \cdot \frac{90}{100} = \frac{18}{1000}$

This produces the following ratio: $\frac{p(X|Y) \cdot p(Y)}{p(X)} =$

$$= \frac{\frac{18}{1000} \cdot \frac{2}{100}}{\frac{116}{1000}} = \frac{18 \cdot 2 \cdot 1000}{1000 \cdot 100 \cdot 116} = \frac{36}{1116} = 0.0031$$

We have taken the formulae from books on stochastic.

The subject felt more confident in her overall precise solution (producing a confidence rating of 3), although even the magnitude of the answer is completely out of range due to the incorrect decoding of $p(X|Y)$. The example may be regarded as a prototype for a one–sided mechanistic analytic procedure which is not accompanied by a sensible understanding through feeling and empathy. Obviously, the statement "The probability of a defective screen is 0.0031" lacks any (intuitive) meaning, although the same subject easily outlined a neat rough draft calculation on the exact 'normative solution' immediately before.

In the course of the trials: From intuition to analysis and vice versa? If individuals deal with a problem repeatedly, starting with a pure analytic (or intuitive) mode of thought, one might suspect the one–sided analytic procedure to be supplemented by the opposite intuitive (or analytic) reasoning. However, based on the analysis of the Majority rating, no evidence could be found to support this conjecture.

CHAPTER 4.6.

Intuition provides middling responses against an idiosyncratic background. One of the female Social Science students provided justifications with increasing intuitiveness scores ending up on one extreme of the analyticity – intuitiveness scale. She dealt with the probability judgment in a double Social Judgment Frame and produced correct reproductions.

1. First I have answered the question 2, as it is completely clear to me that my estimation is built on unsafe ground. On the one hand, this is due to my ignorance and indifference toward such music forecasts, on the other hand, I do not have any feeling for "hitproneness" or not, and that is why I cannot judge people's judgments.
2. The answer came about through a mathematical weighing. On the one hand: a) 10% of all introduced titles become hits.
   b) The studio guest's prediction seems to be correct in most cases (80%), mean: 50%.

This justification received three intuitive scores (B, C, G), the other dimensions were not rated. Apparently, the 50% response had been nonalgebraically fixed when referring to subjective ignorance and a nonalgebraic middling operation on the offered pieces of information. The reasons for this middling strategy were not given, but it seems likely that no analytic penetration into the dependency relation was carried out. The subject stuck to her response on the second trial. The 'mathematical examinations' were completely repressed, and personal arguments about the hit parade were outlined. The second trial justification reads as follows:

Considerations:
– Which titles become hits and why?
– Which conditions have to be fulfilled in order that a title becomes a hit?
– What are the psychological effects of titles which become hits. When a listener chooses a title or not, on what does he base his judgment? – How does the music fashion change, **who** changes it, where are the impulses for shifts in judging music? (indications)
– What sort of listeners listen to the hit parade?
– How important is the hit parade for such listeners?
  (singing along, getting through the day, early morning shocker, etc.)

These considerations have not directly guided my judgment. But they make me notice that, as it is hard for me to clarify my considerations, I cannot produce a safe estimation. Hence, I stick to my mathematical mean.

When asked to work on the Hit Parade problem at home, some more background is revealed for this subject's dealing with base – rate problems.

At home, 8.20 p.m.

First about this fan. He recorded data about how hits came about. What are his/her motives? Which ideas did **he/she** have when doing this? All that he can do is "figure" out pure numbers, but the real background to this

remains hidden. By background I mean all the considerations that came to me when going through the text in the second round.
Furthermore:
Based on which knowledge about the lawfulness (that makes a pop song become a hit), i.e., how songs are composed/texted etc.? What must a singer look like to "supply" a hit? Which is more essential, the text, the music, or the singer? What aspirations do the people have who are singing songs? Their own, one that is dictated to them, or none at all? What I cannot conceal when making all these considerations, is:
During these ideas, again and again another idea jumps in, I am making these considerations as I have been told to do them! I cannot really describe them as being very "intuitive".
Hit parade – a fossil from my youth, when I crept into the cold cellar (where the TV was kept), with blanket and sandwich, when I admired the super men and women, who were singing so self–confidently – and I was still so small ... Such people (adolescents) are presumably the audience or the TV watchers of today, I am thinking.
What sort of person is this studio guest. If he makes 80% correct predictions, he must be an insider, mustn't he? That means, he is able to judge – in a for me inexplicable manner – all the factors which cause a song to become a hit. Or he just has the right "feeling" and does not need any factors and no internalizing of constraints. – He's simply got it under his skin.
Curious world.
One should conduct an experiment on this ...
Hit parade – for me it is furthermore tied to 'lower music' – as opposed to classical music or jazz, for instance – nevertheless, I propose that in this thing no superficial mechanisms are decisive, but more subtle factors, which I cannot grasp.
I rate the probability that the title will be chosen as a hit by the listeners at 50%.

We feel that such protocols not only make for enjoyable reading. Argumentations of this type have been repeatedly observed. They are part of the reality of experiments in probability judgment and also of stochastic thinking. If we wish to construct explanatory and descriptive models, they should encompass all the inferential processes revealed, and not try to ignore these types of cognitive activity. The modes of thought investigated above represent the beginning. A possible theoretical framework that specifies and elaborates the constructs of modes of thought will be presented in the next chapter.

CHAPTER 4.7.

4.7. AN INVESTIGATION INTO EXPERTS'S (SOPHISTICATED DECISION MAKER'S) BEHAVIOR IN BASE−RATE PROBLEMS

Can experts in stochastic master base−rate problems if they behave naively. In other words, do they produce fewer diagnosticity and more middling responses that are close to the 'normative solution' than nonexperts? Does their professional competence help if they are dealing with base−rate problems in a nonprofessional way? How do expert statisticians cope with base−rate problems if they are processing in the intuitive mode? Are there any differences in problem understanding, in problem representation, or solution strategies (cf. CHI, GLASER, & REES, 1982; GLASER, 1984)? These are the key questions which are investigated in Experiment D.

Like many issues of research into decision making, the answers to these questions are controversially disputed. In order to provide some insight into the tension inherent in the 'expertise discussion' on the 'base−rate fallacy', the author wants to report some concrete (anecdotal) evidence.

In discussions following lectures on the 'base−rate fallacy', two arguments are standard. First, subjects' generally weak performance in the experiment is attributed to their naivety. Second, doubt is cast on the meaning of the findings because of the artificial character of the experiments and the tasks used.

The author has often experienced that people who have some formal and mathematical competence (but lack experience in experimental psychological research) often laugh when students' response distributions on the (relatively mathematically trivial) base−rate problem are reported. Furthermore, the analogy "What psychologists are doing with the subjects is the same as if one would ask a sample of nondrivers without driving licences and any driving competence to ride in crowded traffic. No one should wonder if they are bumping" is sometimes expressed. Even the results of TVERSKY and KAHNEMAN's, 1971, experiment on the law of great/small numbers carried out with "sophisticated psychologists" (cf. TVERSKY & KAHNEMAN, 1974) is often questioned, as doubt is cast on the competence of these (nonmathematical) statisticians.

Against this background, the following Experiment D which was run with a sample of experts in stochastic might help to clarify the validity of these arguments. Furthermore, this experiment will seek to gain nonpsychologist experts' judgments on the worth and importance of experiments using variants of the Cab problem like the students' experiments (Experiments B and C), after the expert has personally suffered participation.

There is some tradition of expert studies in individual decision and problem solving research. For instance, the studies by CLARKSON, 1962, on portfolio selection, DE GROOT's, 1965, seminal work on chess playing experts and novices' solving of physical problems (cf. LARKIN, McDERMOTT, SIMON, & SIMON, 1980), or the research on probability calibration by physicians and meterologists (cf. LICHTENSTEIN, FISCHHOFF, & PHILLIPS, 1980) have shown the potentials but also the limitations (cf. SELTEN, 1983) of human decision making experts. Within the tradition of problem solving and information processing, it has often been emphasized that experts apply different knowledge and use other concepts than laymen if they have to deal with "semantically rich domains" (cf. SIMON, 1979, p. 362; CHI, GLASER, & REES, 1982; SCHOENFELD, 1983) and difficult problems.

Before we turn to the method section, we will clarify in which sense university professors in statistics are considered to be experts. We will introduce the distinction between **experienced** and **sophisticated** decision makers. By experienced decision makers we mean individuals who either have performed a certain decision repeatedly or have experienced the information, e.g., have observed the concrete history of the base−rate information (cf. CHRISTENSEN−SZALANSKI & BUSHYHEAD, 1981). This definition of expertise is rather narrow and domain−specific (cf. SCHOENFELD, 1983). For instance, it allows for professional mathematicians, say topologists, to serve as novices in another field of mathematics, such as number theory. By sophisticated decision makers, we mean individuals who possess the ability or knowledge to cope with a situation or find an adequate solution for a problem.

Without any doubt − when working on the Hit Parade problem and other base−rate problems − university professors in statistics should be considered to be **sophisticated** decision makers, as a handling of the formal structure and the type of word problem is on a low level of their professional activity. But not all the professors can be regarded as experienced decision makers, as many of them are neither fans of the hit parade, nor is the iterated coping with base−rate **word** problems in a social judgment cover part of the standard statistics curriculum. In some respects, however, university professors in statistics are **the** experts for word versions of the base−rate problems, as abstract, formal, and/or verbalized problems are an object of their profession.

4.8. EXPERIMENT D

4.8.1. SUBJECTS

The subjects were ten scientists from Bielefeld University. All of the professors were involved with fields of disciplines in which the probability concept is essential. Hence, the subjects could be called sophisticated decision makers. It should be mentioned that not all experts were specialists in teaching courses on probability and statistics. Theoretical physicists, for instance, rely heavily on the probability concept in their theories, but usually integrate instruction on probability into courses about quantum mechanics, etc. The sample consisted of four mathematicians, three theoretical physicists, and three social scientists. The professional status of the experts varied. There were four full professors, four associate professors, and two "Akademische Räte" (i.e., lecturers or research associates, who are members of the permanent staff of the university).

4.8.2. PROCEDURE

The expert sample was gathered by consulting students, secretaries, or scientists about the expertise of members of the staff in respect to the objectives of the study. The experts were visited by the author and asked whether they would participate in an expert study on stochastic thinking divided into three sections. There was only one refusal from an educational scientist. First, they should work on some stories and assess judgments of probability and percentages (reported here), second, their understanding of the probability concept would be explored by an interview, and third, the experts should evaluate the relevance and the procedure of the experiment which had also previously been conducted with a student sample in a similar manner.

The individual sessions were conducted in the Psychology Department. The experts were given a booklet which described the former student experiment, especially the features of the Experimental Framing (see Chapter 4.5.2.). Then the subjects were instructed to work on the Eye Color and Hit Parade problems (see Chapter 4.5.2.) and to produce an 'intuitive' estimation or probability judgment, but were asked not to look directly for a formal solution, although one might possibly exist. Subjects were assured that their experimental behavior would be handled anonymously. Furthermore, they were asked to behave naively. But it was also explicitly stressed in the instructions that naive should on no account mean that they should try to shed their (expert) knowledge, but rather, that they should try to start from the experimental conditions as described in the booklet. On the last pages of the booklet, the Hit Parade problem was presented once again, and the experts were asked to sketch a formal solution which they judged to be correct.

4.8.3. RESULTS

Text understanding. Although the experts were not asked to reformulate the problems and questions in their own words, some information was gained on the text comprehension. One cue is given by the written justifications, another by a short discussion about their procedures and their understanding of the base—rate problem. Three subjects definitely started from a matching probability interpretation of the diagnosticity, and six subjects started from the 'standard' interpretation in both trials. An idiosyncratic interpretation was given by the tenth subject. This subject did not record the diagnosticity specific to the outcome of the events, but started with a .8 probability for the studio guests' "Hit" judgment, and a .1 probability for the listeners "Hit" judgment. The missing parameters were then subjectively assessed with $p(\text{Hit}|\text{"Hit"}) = .5$.

Modes of thought. Experts' justifications in the 'intuitive' condition were submitted to an analytic vs. intuitive rating by the author; six subjects were judged to process intuitively, three analytically, and one subject could not be classified. One of the three analytic subjects refused to respond intuitively, as he knew that such an answer would be "cogently" incorrect and hence inappropriate. This subject was the only one who produced an algebraic solution in the 'intuitive' experimental condition. A second analytic subject started from a matching probability understanding which he carefully checked in his justification. The third precisely analyzed the impact and the interaction of the information while having BAYES' theorem in mind, but avoided numerical calculations. The mean of intuitive features across all subjects in the first trial was 4.2, and the mean of analytic features 2.7.

In the analytical condition (the second trial), it was judged that all experts processed analytically. Two subjects became irritated, presumably because of the unusual test procedure, and did not manage to produce a (analytic) second response. However, these subjects continued to cooperate in the subsequent sections of the experiment.

Response distributions: The modal response in the intuitive trial was the diagnosticity (cf. Table 4.12), and the mean normalized deviation $|d_2|$ was .37. The modal response on the second trial was the 'normative solution'. Only diagnosticity and 'normative' responses were produced during the second trial. Relative to the 'normative solution', four subjects improved their performance on the second trial, while one subject did worse. In the intuitive trial, this expert started with a matching probability understanding, but then performed a subjective reassessment of the matching probability (instead of .8, he considered .7 to be intuitively appropriate!). In the analytic trial, he returned to the information in the problem and produced a diagnosticity response. In the first intuitive trial, the two other experts who started with a matching probability (i.e., the diagnosticity) understanding consequently produced a diagnosticity response, whereas three out of the remaining seven, who started with the intended text understanding, produced the diagnosticity.

Table 4.12.: Parameters of experts' reponse distribution

	First trial (intuitive) N=10	Second trial (analytic) N=8*
$|d_2|$ error measure	37	18
'Normative solutions' in %	10	62
Diagnosticity responses in %	50	38
Middling responses in %	40	–

*) In the second trial two experts did not complete the task, presumably because of experimental stress.

The experts' evaluation of the experiment. Although not all subjects managed to cope with the problem introduced, there was a strong overall agreement on the importance of the experiment and the meaningfulness of the experimental tasks. In the postexperimental interview, all but one expert agreed on the 'standard normative solution' for the Hit Parade problem as the only appropriate solution. The opinion of the subject who disagreed will be discussed in the next section.

4.9. DISCUSSION OF EXPERIMENT D'S RESULTS

University professors, who were experts in stochastic, participated in an individual session experiment. They were informed about the different experimental framings under which students had participated, and were asked to place themselves in a Social Judgment Frame and to generate an intuitive judgment.

Experts' intuitive judgments produce a lot of diagnosticity responses. The distribution of the experts' intuitive responses was similar to that of postgraduate students in the Hit Parade problem or other versions of the Cab problem (cf. Table 2.3.). We think that the high percentage of diagnosticity responses has to be regarded as a characteristic of more highly educated samples (cf. the discussion of Experiment A's results, Chapter 2.7). Although the sample size was only ten subjects, it should be mentioned that our professional sample produced the highest percentage of diagnosticity responses to be observed on the Hit Parade problem in our studies. Unlike the student sample, the majority of the diagnosticity responses were based on intuitive strategies (see below), and only one out of five was clearly the result of an analytic procedure (starting form a different text understanding).

Experts' (middling) responses are systematically biased. Experts' intuitive judgments are nonrandom and systematically biased. Given the specific order of base−rates and additional information, no expert underestimated the 'normative solution'. Thus the knowledge about the nonlinear relationship between base−rate and diagnosticity obviously had not been internalized, and the middling responses showed a systematic bias toward an overestimation of the 'normative solution'.

Inverting the diagnosticity response is not only a student error. Like some of the students, several of the experts commenced with an understanding of the inverted conditional probability. Postexperimental interviews showed that these inversions had various causes, for instance, the text was not read all the way through, or the meaning was not completely penetrated. All but one expert, however, agreed on the 'standard normative solution' for the Hit Parade problem when asked which solution should be regarded as the adequate one. The expert who disagreed regarded the diagnosticity information, interpreted as general matching probability, to be an adequate answer. He justified this opinion by pointing out the everyday character of the problem, and indicated that the generally valid common sense meaning of the wording: "The studio guest made 80% correct predictions, both for the songs that became accepted into the hit parade and for those that did not" expresses the probability p("Hit"|Hit) = .8. Obviously, this expert possesses two language codes. According to him, no base–rate considerations can be taken into account in the Hit Parade story, as the case–specific validity of the matching probability can not be altered by the consideration of base–rates within everyday language. Hence, the second part of the above Hit Parade text, the explanation of the case–specific meaning, was irrelevant for him. When other contexts such as the TV problem were presented to this subject in the postexperimental interview, he agreed upon the Bayesian revision.

Judgmental heuristics can be found in experts' justifications. University professors, like students, show a multitude of strategies in assessing probability judgments. Contrary to the above student justifications (see Experiments B or C), the causal scheme and the representativeness heuristic seemed to guide some of the experts' judgments. A causal relationship between the studio guest's prediction and the listeners' choice is seen in the following argumentation by a natural scientist.

> The listeners **always** hear the studio guest's prediction, hence they will make the same decision by this random title as they have made before.

The next justification clearly relied on the representativeness argument.

> I mainly include the fan's experience. My uncertainty is due to the following reasons. How representative are the specific studio guests?

And post hoc he described his strategy:

> I identified myself with the studio guest and considered myself to be representative.

Experts' analytic responses: Either the 'normative solution', the diagnosticity, or none. Experts know that it is possible to determine a meaningful mathematical solution, and they know that they are capable of understanding the principles behind such solutions. This is presumably why they exclusively responded with either the 'normative solution' or the diagnosticity, or they did not give any answer. No wild calculations such as had been observed in student samples were to be found here. A plausible explanation for this difference has been provided in CHI, GLASER, and REES', 1982, survey on "Expertise in Problem Solving". Process studies on physical problem solving show that novices' knowledge and strategies are organized around the objects or explicit cues, whereas experts organize their knowledge around fundamental principles. Furthermore, novices consider a greater number of cues as being essential for the solution of a problem (cf. CHI et al., 1982, p. 65), and student novices rarely see the underlying principles. This may be why many of them start from the explicit numbers given, and then try to combine these in a rather disorganized and unsophisticated bottom−up manner. Experts, however, see the principles behind the base−rate problem and recognize the inappropriateness of such wild calculations. Hence, they either respond with the 'normative solution' (if they are given enough time, a suitable problem solving frame, and the necessary tools, that means paper and pencil), produce a biased estimation (if the tools are missing), or simply stick to the diagnosticity information, which seems to have some intuitive appeal for experts.

PART II

5. STOCHASTIC THINKING, MODES OF THOUGHT, AND A FRAMEWORK FOR THE PROCESS AND STRUCTURE OF HUMAN INFORMATION PROCESSING

"The previous contributions have clarified that we cannot deal primarily with the question of quantitatively measuring knowledge, but more that we should record moments of the structural state of the knowledge process at a specific point in time and then compare points over time. The recording process must be guided by theories on the representation of knowledge."
(MANDL, SPADA, ALBERT, DÖRNER, & MÖBUS, 1984, p. 34, our translation)

This quotation was originally formulated within proposals for a ten year key program on the psychology of knowledge (Wissenspsychologie) directed by the German science foundation (DFG). Yet the calls for, and the challenge of a qualitative, structural, and concept−driven modeling of cognitive processes also reveals the indispensible necessity for research into decision making under uncertainty or stochastic thinking. In the field of probability judgments, this necessity has been particularly highlighted by the research and findings on the 'base−rate fallacy' which have been described or developed in the previous chapters.

The introduction of different modes of thought can to some extent be regarded as a **third step** toward an explanation of cognitive strategies in probability judgments. We consider the **first step** to have been the introduction of various unrelated judgmental heuristics such as the representativeness heuristic (KAHNEMAN & TVERSKY, 1973), the availability heuristic (TVERSKY & KAHNEMAN, 1973; NISBETT & ROSS, 1980), the causal schema (AJZEN, 1977; TVERSKY & KAHNEMAN, 1979), etc. (cf. Chapter 2.3.). The **second step** consisted in the introduction of information weights (cf. Chapter 2.4.) in order to provide a rough model, that is the individualized normative solution, of the internal cognitive representation for the reassessment of the given probabilities, and also for intermediate answers (i.e., an answer that falls between the diagnosticity and the 'normative solution'). The formation of the model was based on differences in the individual interpretations of the problem, that in turn, partially relate to different variants of the probability concept; for instance, subjective or logical probabilities. However,

CHAPTER 5.   143

the analysis of the answer and solution protocols (cf. Chapter 3.) not only revealed that a multitude of cognitive strategies were applied by subjects, but also that the strategies applied appeared to differ in quality. The complementarity of the **intuitive** and the **analytic** modes of thought was introduced to conceptualize these differences. A definition of the modes of thought was originally introduced by lists of attributes or features. These lists of features also provided the basis for the classification of cognitive strategies which was applied to the written justifications or procedure protocols that subjects produced when assessing probability judgments. It was possible to test and confirm the interrater reliability and the construct validity of the modes of thought in stochastic thinking in two independent experiments (Experiments B and C). Furthermore, we obtained valuable information on the effects that framing and career socialization have on cognitive strategies (cf. Chapter 4.).

In our opinion, the analysis of written justifications and the introduction of modes of thought made it possible to gain important insights into the nature of information processing in probability judgments. However, as we already formulated some years ago when criticizing the diverging and theoretically unlinked judgmental heuristics, "**A better understanding of the total cognitive activity will only be obtained when the specific form of knowledge storage (the knowledge base), the acquisition of information, and information processing are clarified in one model**" (SCHOLZ, 1981, p. 11). The modes of thought are not such a model, and to date, there are only rough model schemes (like that of HOGARTH, 1980; HOGARTH & MAKRIDAKIS, 1981) or phase models of stochastic thinking (cf. SCHOLZ, 1981) that do not provide a satisfactory basis for answering, for example, the following questions:

— In what ways are the specific knowledge base, the factual knowledge, and the meaning of concepts or previously learned algorithms involved in probability judgments?
— When and why are specific heuristics or rules applied, informations weighted or subjectively reinterpreted, etc.?
— When and why is which mode of thought activated?
— When is the processing of information interrupted or halted, and when is a decision made or a response given?
— How are the various heuristics tied to the different modes of thought?
— Which response distributions are produced by the various heuristics and different modes of thought?
— And last but not least, how may the different approaches (i.e., the different

"steps" described above) to a conceptualization of probability judgments and the various findings from Experiments A to D be theoretically integrated?

5.1. A MODEL OR FRAMEWORK FOR THE PROCESS AND STRUCTURE OF HUMAN INFORMATION PROCESSING

The structural model for the acquisition and processing of information presented in this report attempts to provide a framework within which the above questions may be answered. We shall try to work out, on the one hand, how the modes of thought and the various judgmental heuristics in probability judgments may be conceptualized within such a framework, and on the other hand, point out what knowledge about the units and relations of the model can be gained from the introduction of the modes of thought and the various findings of our experiments. Finally, we shall consider what new perspectives for an understanding of the process of stochastic thinking (e.g., the impact of the role of emotions on the processing of information) may be gained by the introduction of this framework. We will particularly discuss how the framework can be used to grasp, to conceptualize, and to experimentally investigate the specifics of probability judgments and stochastic thinking.

Before presenting this framework, it seems advisable to add some remarks on the use of the terms **model**, **theory**, and **framework** in the present chapter. Within our understanding, a **framework** is a general pool of constructs and concepts for understanding a certain domain, yet it is not tightly enough organized to derive precise prediction (cf. ANDERSON, 1983, p. 12). In psychology, the terms theory and model are often sloppily defined. Usually, the term theory is understood as a precise deductive system which leads to accurate explanations of phenomena, whereas models usually are regarded simply as rather domain–specific theories. In this connotation, the term **model** is an adequate expression for the individualized normative solution, which has been introduced in Chapter 2., whereas that which will be outlined in the following should be strictly denoted as a conceptual framework of the cognitive activity in stochastic thinking. However, as in some of the major references given in the subsequent framework the term model is used (cf. HUSSY, 1983; KLIX, 1980), we will at times maintain this usage although it does not meet the above definition. The author also wants to note at this point that he considers the general system approach, but not the (special) computer analogy, to be the basis for the subsequent framework. Nevertheless, the framework will be presented in such a manner that it may provide a basis for structuring computer simulations of probability judgments in certain well−defined situations.

5.1.1. AN INTRODUCTION TO THE COMPONENTS OF THE FRAMEWORK

Alongside the above−mentioned work of HOGARTH, 1980, and SCHOLZ, 1981, the framework is above all based on the work of DÖRNER, 1976, 1979, 1981; HUSSY's, 1984, structure and process model of complex human information processing, and KLIX', 1980, differentiated model of information circulation. It will also be pointed out that some analogies can be made to ANDERSON's ACT−star−model (cf. ANDERSON, 1982, 1983). We will point out the references to these approaches and the tradition of the proposed framework in Chapter 5.1.3.

Stochastic thinking is thinking. This means, on the one hand, that the same components that generate other thinking processes also provide the basis for stochastic thinking. On the other hand, the contents, the concepts, and also to some extent the functions and cognitive operations, differentiate stochastic thinking from other forms of thinking such as logical (cf. JOHNSON−LAIRD, 1984), geometrical (cf. LAUGWITZ, 1979), functional (cf. v. HARTEN et al. 1985), and recursive thinking (cf. HAUSMANN, 1985).

Some methodological remarks on the relation between thinking, stochastic thinking, and probability judgments. In order to clarify the relation between the general and the specific, between thinking, stochastic thinking, and probability judgments in base−rate problems, we wish at this stage to comment on the area covered by this text and cn the different levels of our analysis. We should like to remind the reader that the actual object of our studies is the conceptualization of cognitive strategies in stochastic thinking, while our experiments are directed at a special, though central, paradigm, namely, probability judgments in base−rate problems. If we wish to use the experimental results to generate predictions about stochastic thinking, then it is necessary to generalize; that means, proceed from the specific to the general.

The opposite approach, from the general to the specific, is used in the presentation and discussion of the framework. In the introductory description of the components, we will only concentrate on marking the general functions that are referred to by these components. For each individual component, however, we will introduce a row of comments and examples in order to illustrate specific cognitive activities in stochastic thinking. We will point out in a precise and specific manner what knowledge, which operations, evaluations and goal setting, and what filtering processes are involved during stochastic thinking. This will show how all the components of the framework are **necessary** in order to be able to record and describe the view of cognitive activity during stochastic thinking that has been developed in Chapters 2. to 4. The examples and comments will also clearly show that, at the present time, we only have varying degrees of knowledge about the individual components. To some extent, we can support statements with our experimental findings, and sometimes we will introduce other relevant findings from investigations into stochastic thinking. It is, however, clear that the present framework has at best a descriptive value if it is to be used backwardly—directed toward the experimental effects reported in Chapters 2. to 4. The value of the framework as a basis for formulating specific hypotheses on stochastic thinking, and the extent to which it fulfills an explanatory function will be considered in the final section of this book (Chapter 5.4.).

We want to conclude our methodological remarks on the framework by noting in advance that the components of the proposed model are regarded as theoretical constructs of functional units but not as physiological or organic elements.

The components of the framework. Figure 5.1. is a framed flow chart with nine elements or units of information processing that are interconnected through twenty internal relations. Additionally, it contains four external relations that are placed between the internal elements and the output or the environment. We will begin by explaining the information processing units.

(1) **Sensory System** (S): The perception of information and the first internal recording represent the immediate direct interface between the human information processing system and the stimuli source.

    Examples/Comments
      a) Although it is true that most stochastic or stochastically interpreted information is received through visual or auditive channels, the direct stimuli may be received through other sense organs. For example, in

---

Reading note: In order to facilitate an understanding of the framework, its components, relations, and functions, the reader may unfold the last page of the volume and view Figure 5.1. while studying this chapter without having to continually turn the pages.

CHAPTER 5.1.1.

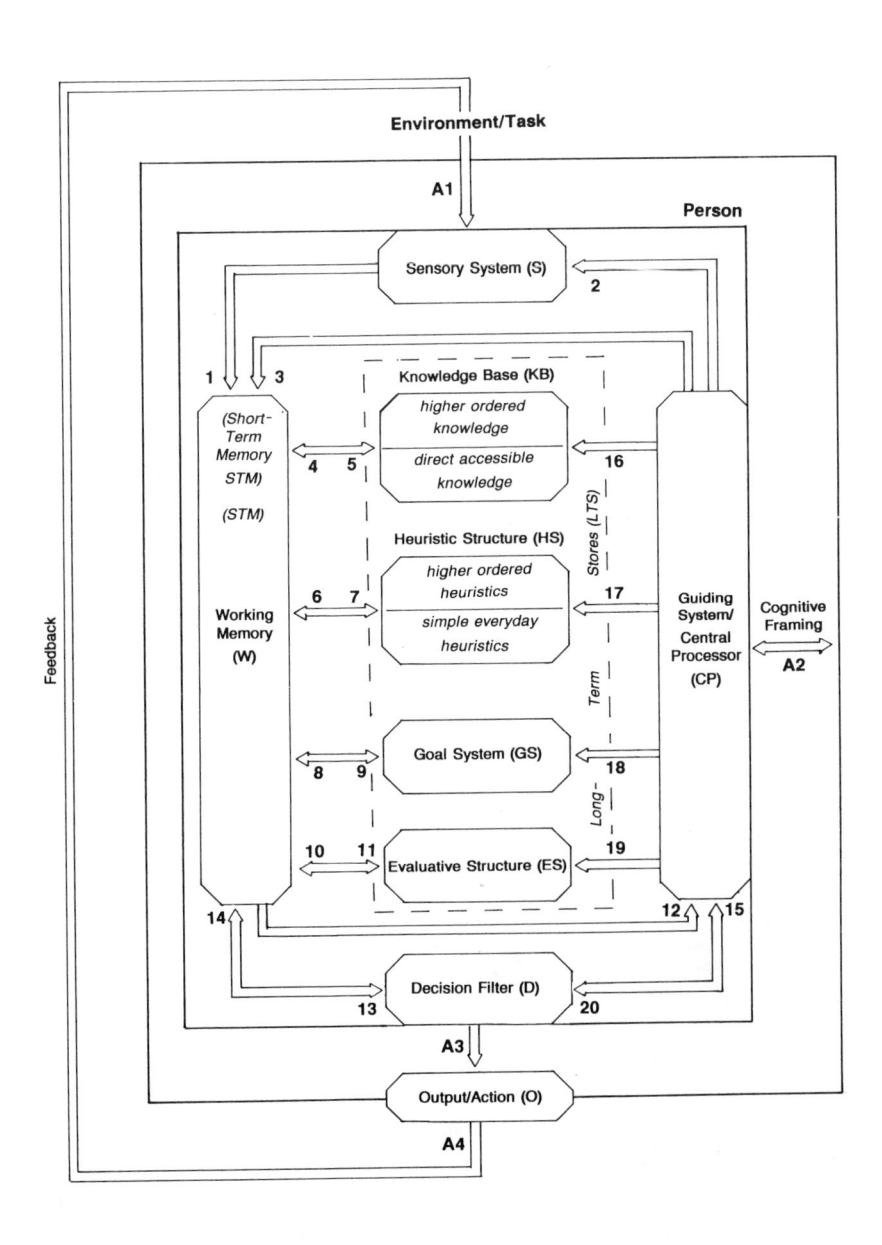

Figure 5.1.: A framework of the structure and process of information processing: The general model

the quality control and blending of tea, cognac, whisky, tobacco, etc., the senses of taste and smell are the primary receptors on which a sophisticated stochastic decision process is built.

b) As the initial phase of information encoding is not of central interest in the context of the experimental analyses reported above, we did not deal in detail with the very beginning of information acquisition, but rather presupposed that the text had been acquired on a literal level. It is well-known that not all incoming stimuli that bring about neural excitation enter the working or short-term memory. Perception is by no means a simple passive or mechanistic process (cf. BROADBENT, 1971; BJORK, 1975; NEISSER, 1976; KLIX, 1980; HUSSY, 1983; SEILER, 1984), and there is an abundance of empirical evidence that points to the existence of a filtering process occuring during sensory registering. Whether and how this filtering may affect the process of stochastic thinking is a question which has to be dealt with in future studies. The potential impact of selective filtering on probability judgments in the early stages of information acquisition has been repeatedly brought up in the previous chapters.

For instance, within the classical definition of the **availability** heuristic (cf. NISBETT & ROSS, 1980; or Chapter 2.3.), a differential ease in the encoding of information input is assumed. Hypothesis 3 in Chapter 2.5. on the differential effect of numerically displayed medium and extreme base-rates was partially founded on this assumption. However, the results from Experiment B did not reveal clear evidence to support the hypothesis on the higher availability of extreme (low) base-rates.

c) We will give another example on the availability heuristic which is a much discussed phenomenon in stochastic thinking (cf. Chapter 2.3.). One special case or variant of availability is the dependence of the speed of information-input-decoding on the frequency or familiarity of the stimulus being recorded (cf., for example, ROTHE, SEIFERT, & TIMPE, 1980). A first type of preadjustment could result from assuming a varying sensitivity or receptivity for (elementary physical) stimuli in the input register that could be modified by the frequency of previously received stimuli. Such a concept has been modeled in signal detection theory (cf. GREEN & SWETS, 1966) and may lead to systematic biases in (frequentistic) probability judgments.

(2) **Working Memory** (W): Within the present framework, the Working Memory is the main operating unit and is responsible for various activities in the course of information processing. For instance, this unit is assumed to be accountable for the temporary storage of processed (i.e., encoded or retrieved) information. This part of the Working Memory will be called Short-Term Memory (STM). Besides handling the encoded or retrieved information, the Working Memory has to deal with specific goals, and the heuristics and evaluative operators activated by the current problem. Based on the concrete

loading with, for example, information and goals, the Working Memory is considered to be the unit in which regulation or search commands are formulated or processed, for instance, for knowledge retrieval or for suitable problem solving heuristics.

Examples/Comments

The first two examples both illustrate the Working Memory's activites in the elementary process of frequency or frequentistic probability estimation. We will present two examples of this in order to reveal the different activites in one and the same task.

a) When, for example, an individual explicitly deals with the estimation of the frequency of an event in which she or he is involved, his Working Memory could be employed to systematically register the frequencies of occurence and nonoccurence in order to consecutively determine the relative frequencies. Optimally, (statistically) sufficient information should be retained from which the total number of events and the frequency of their occurence can be determined. Both these values must be registered and available for revision through future observed events. In these situations, the processing rules or heuristics could be known elementary algebraic operations. When this is the case, we are dealing with a simple analytic procedure for determining frequencies that can be interpreted in a frequentistic probability frame.

b) A subject may proceed in a less systematic manner. For example, he may simply recall a few past events into her Working Memory. Then, when either actual easily accessible events or the recording capacity are exhausted, he may proceed by determining what seems to him to be an adequate frequency, basing this decision on the directly available loading. We would describe this as an intuitive frequency estimation. According to NISBETT and BORGIDA, 1975; NISBETT and ROSS, 1980; and LICHTENSTEIN, SLOVIC, FISCHOFF, LAYMAN, and COMBS, 1978, such intuitive frequency estimations are prone to systematic biases because of availability effects (cf. Chapter 2.3.).

c) We will present a third example to illustrate both the complex functions of the Working Memory in stochastic thinking and the value of the framework for an understanding of information processing in everyday situations. A car refuses to start, and there are several sources that could be considered responsible for the defect. If, for example, two different sources (such as the fuel system and the electrics are suspected, and a controlled series of routine tests (is there petrol in the tank; are the spark plugs functioning correctly) has been carried out, a fault analysis that requires the expenditure of much work and material, with roughly the same amount for both sources, might become necessary to investigate each of the sources. Clearly, the decision what to do next should depend on the likelihood of being defect which is attributed to each source. One might think about the situation and then eventually process the question as to which of the two sources is the most probable source of the defect. In this situation, the Working Memory will be busy consulting and actualizing the Knowledge Base in a search for indicators that point to one of the two sources. Further, one would bring to mind any further indicators that could be investigated as possible defect sources. In order to then decide whether to change the fuel pump or fit a new distributor, one needs a decision rule. One possible decision heuristic would consist in determining the relationship between positive indicators investigated and source – specific indicators investigated for each source. (This decision rule has actually been applied by subjects in Experiment B, cf. Table 3.8., when working on the Motor base – rate problem). Then one could begin the actual physical work by replacing the part with the larger relative proportion of positive indicators. Of course, such a procedure not only takes a lot of effort and places heavy demands on Working Memory capacity, but it is also relatively unreliable, as one can hardly assume a complete availability of defect sources in this real world example (cf. FISCHOFF, SLOVIC, & LICHTENSTEIN, 1978). Empirical rules of thumb, such as one does that which was successful the last time one experienced a comparable conflict, may well be consulted instead of the above – mentioned complicated procedure.

d) The way in which search and regulation commands are incorporated within the Working Memory may be illustrated by Experiment D (cf. Chapter 4.8.2.). The subjects in this experiment, the sophisticated statisticians, were fully informed about the objectives of the experiment. However, they were asked in the instructions not to look for a formal solution, although in the Hit Parade problem such a solution existed, and experts in general usually immediately realized this. Within the terminology of the present framework, this question may be considered as a request to restrict the search space in the retrieval of knowledge and heuristics.

e) Introspection may be considered as a special search command of the Working Memory in the sense described above in Example d). Some authors, however, postulate a differential ability of introspection in respect to knowledge and heuristics. According to ANDERSON, 1983, or ALLWOOD, 1985, the declarative knowledge may be inspected, but not the procedural knowledge. Our view differs from this and will be outlined below.

(3) **Knowledge Base** (KB): The Knowledge Base, otherwise known as factual knowledge, knowledge storage, declarative memory, or epistemic structure, in our view contains that knowledge which is internalized and permanently retrievable (cf. DÖRNER, 1976, 1979; KLIX, 1980; ANDERSON, 1983; HUSSY, 1983, 1984). In accordance with the viewpoints of recent memory research, we assume that concepts and cognitive routines

was found in the experts' intuitive responses, whereas student samples produced the diagnosticity response predominantly in the analytic mode (cf. **Experiments B and C**). According to Assumption 2 within the text on the Knowledge Base, we propose a distinction between direct accessible knowledge and higher ordered knowledge. Furthermore, we noted above (cf. Chapter 4.9. when referring to CHI et al., 1982) that the experts or sophisticated decision makers' Knowledge Base is organized around fundamental principles or functional properties, whereas the novices' Knowledge Base is organized around facts and/or isolated concepts. A potential explanation of the different inferential bases of the students and the professors' diagnosticity responses is given by assuming different Knowledge Bases for these two samples. It seems to be plausible that the professors' Knowledge Base is structured in such a manner that, even when processing intuitively, they immediately realize and know that the Hit Parade problem does have (a numerical and) a unique solution. As they were asked in the instructions not to apply formal calculations to produce a response, they might have answered with the probability estimation which they superficially regarded as being most plausible (intuitive numerical solution). When student samples are applying their direct accessible knowledge, they much less frequently realize or recognize the mathematical structure in the Hit Parade problem, and hence their intuitive judgment leads to middling responses. However, if they are processing analytically, they look more closely at the problem, classify it as a problem with a mathematical solution, and respond with the diagnosticity (as their heuristic tools are not sufficiently developed, see below).

f) The need to consider the content of the individuals' Knowledge Base has been convincingly demonstrated by R. MAY, 1986a, b. The object of her studies was the individual's subjective probability ratings in classical calibration tasks. In these tasks subjects have to rate their confidence in their answers on problems of the type: Which is further north: New York or Rome. In her careful analyses, MAY revealed that the subjects' overconfident responses on these items may be explained by the subjects' actual knowledge about the geographical latitude of North America and Europe. That which has been traditionally but superficially explained as the subjects' fallacious proneness to be more confident dissolves if the actual knowledge on the item on which the the probability statement is based is taken into account. In her conclusion she argues that overconfidence is more a property of the misleadingness of the task (cf. SCHOLZ' definition of paralogisms, 1981, p. 10 – 16) than of a fundamental bias in subjects response behavior.

(4) **The Heuristic Structure (HS):** A heuristic is a problem solving procedure. Problem solving researchers (e.g., DÖRNER, 1979;

LOMPSCHER, 1972) regard heuristics as a sequence of elementary mental operations that are consulted for the solving of problems, and are used to combine and to process the information that is present in the Working Memory. We postulate that the Heuristic Structure consists of a store of operators and decision rules and a set of application rules as to how these operators may be combined.

Although the necessity for a distinction between procedural and factual knowledge has been pointed out repeatedly in theories of knowledge (cf. ANDERSON, 1983; OSWALD & GADENNE, 1984) and psychological memory research (cf. DÖRNER, 1979), there is still relatively little clarity about the nature of the Heuristic Structure. As with the Knowledge Base, we will also discriminate heuristics in the Heuristic Structure according to their degree of accessibility. We will distinguish between **simple everyday heuristics and higher ordered heuristics**. Everyday heuristics are supposed to be fast processing inferential tools that require relatively little storage capacity and are normally acquired in the course of everyday practice. In contrast, higher ordered heuristics usually are gained in the process of tutored activity or formal education. We will introduce below a series of heuristics that should convey our view of this structure.

The problem solving procedures stored in the Heuristic Structure differ from the cognitive routines or the principles that are stored in the Knowledge Base through their greater range of applications, often in the flexibility by which they may be modified or decomposed, and by their status in the course of information processing. Heuristics may operate on knowledge elements to generate new or unavailable knowledge. However, other elements from the individual's memory content, such as heuristics themselves, may also be the object of heuristics. In order to illustrate this, we will first present a nonstochastic example.

Examples/Comments
a) The quadratic equation $(a + b)^2 = a^2 + 2ab + b^2$ may be simply recalled from the Knowledge Base and automatically applied without any know how about the distributive and multiplicative rules that underlie it. If this is the case, usually no elements from the heuristic structure are involved. Oppositely, one may fail to recall this equation, but may know how to handle the distributive law. Within our framework, this operative ability usually will be conceived as a **higher ordered** heuristic (see above), and hence as an element of the Heuristic Structure.

Example a) may also be used to clarify the different status of routines which are retrieved from the Knowledge Base and the Heuristic Structure. If the quadratic equation is simply recalled and directly applied, there will be no barriers, whereas in the second case, the answer is not immediately at the individual's disposal and has to be constructed. According to HUSSY, 1984, such a difference is also essential for the differentiation between tasks and **problems**. But another essential difference between the Knowledge Base and the Heuristic Structure may also be illustrated by this example. New elements of the Knowledge Base may be rather rapidly stored, whereas the acquisition of new heuristics takes a much longer time (cf. ANDERSON, 1983).

The next two illustrations are drawn from choice behavior when making decisions under uncertainty.

b) If there is a choice between two risky alternatives (e.g., two different routes for an alpine trek), one might decide to bring into one's mind different dimensions, on which one could make risk comparisons (e.g., danger from falling rocks, danger of an accident through missing one's footing, crevasses, etc.). A simple rough decision heuristic that one could consult in such a situation is the so-called **majority rule**. This rule of reasoning consists in simply choosing the alternative that is indicated by a majority of those dimensions stored in the Working Memory that have a lower risk probability (of an accident occuring) (cf. ASCHENBRENNER, 1977; SVENSON, 1979; HUBER, 1982). The majority rule seems to be a simple decision procedure that appears to show empirical validity for both intraindividual conflicts and group decisions (if there are individual single-peaked preferences, cf. e.g. CROTT & ZUBER, 1983).

c) A second decision heuristic that could also sometimes be applied in the above, mentioned illustration is the so-called maximin rule. According to this rule, the alternative is chosen that indicates the lowest maximal risk probability in all dimensions (cf. THORNGATE, 1980; HUBER, 1982).

d) Further decision heuristics could be, for example, the conjunctive heuristic or the satisficing principle (SIMON, 1955) that state that the alternative is chosen that satisfies a specific criterion on all dimensions or, in the case of the satisficing principle, dimension-specific criteria that are determined by individual needs (cf. HUBER, 1982).

e) One way of defining representativeness is the determination of the degree of similarity. A typical example of this is the judgment of a person's membership in a certain population (cf. the Tom W. base-rate problem in Chapter 2.1., p. 11). Under special circumstances, the **representativeness heuristic** can also be interpreted as a feature matching majority rule. However, the similarity comparisons appear to be based on more complicated (asymmetric) relations (cf. TVERSKY, 1977). Even when we do not specify the concrete basic principles of similarity comparisons, we regard the representativeness principle (i.e., estimating the probabilities that events belong to a specified population according to their degree of similarity) as a possible element of the Heuristic Structure.

f) We will consider one of the written protocols presented in **Experiment C** in order to exemplify what is meant by higher ordered heuristics. In Chapter 4.6. an example was discussed under the heading "analyticity may lead astray". When interpreting this subject's inferential process, one can see that the subject starts by: (1) separating the various elements of information by fixing the **complete list** of all (conditional) probabilities given in the text (e.g., the diagnosticity information), then (2) **she recalls the requested probability,** (3) **translates this question into mathematical terms,** and finally (4) dissolves this formula when applying theorems of the probability calculus till events may be identified whose probabilites are provided in the text information. Obviously, this subject attacks the base–rate problem using a typical mathematical problem solving strategy analogous to the general heuristics described in POLYA's, 1949, fundamental book "How to solve it", beginning with what is known (1), then what is unknown (2), introduce a suitable notation and restate the problem (3), and carry out calculations step–by–step (4). As we know from mathematics education, this procedure has to be developed through formal education and a lot of practice.

g) In the final example, we once more want to illustrate the difference between **simple everyday heuristics** and **higher ordered heuristics** in the coping with base–rate problems.

Simple everyday heuristics have been documented in the various case examples presented in the previous chapters. In Chapter 4.6., for instance, we discussed an intuitive middling response. When working on the Hit Parade problem, this subject obviously intended to determine the mathematical mean of 10% (the base–rates) and of 80% (the diagnosticity information). The subject's operation resulted in "mean: 50%" which – among other things – definitely reveals the rough, uncontrolled, and unreflected nature of the inferential process.

There are two more examples of such averaging operations in Chapter 3.4. that are also classified as intuitive strategies. According to mathematical criteria, both protocols provide incorrect calculations, yet both document prototypical averaging operations that do have some rationale.

The individualized normative solution is beyond the operative possiblities of simple everyday heuristics. Conclusions or inferences that have been outlined in Examples 2 or 3 of Chapter 2.2., which have also been occasionally observed in real subject's protocols, are prototypes of higher ordered heuristics. Obviously these heuristics are unlikely to be acquired without formal instruction. This is particularly caused by the nonlinear structure of the Bayesian calculus. However, the reassessment of probability parameters and their incorporation into the calculus itself also has to be viewed as a rather subtle analytic procedure.

(5) **The Goal System (GS):** As is shown in Figure 5.1., we differentiate between a Goal System and an Evaluative Structure both of which are responsible for the individual's governing process.

We regard the system that formulates the respective goals, aspirations, and orientations for operations of the Working Memory as a special unit. The meaning of goals is not only relevant for decisions under uncertainty. One example of such goals, namely aspiration levels, plays a decisive role in Kurt LEWIN's action theory (cf. LEWIN, 1926; HOPPE, 1931) and also in bargaining and social conflict theories (cf. SIMON, 1955; SAUERMANN & SELTEN, 1962; SCHOLZ, 1980a; TIETZ, 1983; CROTT, SCHOLZ, KSIENSIK, & POPP, 1983). Within our model of information processing, the Evaluative Structure and the Goal System can be regarded as the most important interface between the cognitive processes (e.g., stochastic thinking) and the emotional and motivational processes (cf. KUHL, 1983a, b; HUSSY, 1984). To a certain extent, we see the Goal System as the unit that provides variable goals and feeds these to the Working Memory, while the Evaluative Structure (see below) contains the operators that make comparative evaluations. Hence the distinction between the Goal Structure and the Evaluative Structure is founded on similar principles as the discrimination between the Knowledge Base and the Heuristic Structure.

The 'dimensionality' of the goals is considerable, and both the organization of the Goal System and the type of goals are crucial for such problem solving processes as may be found, for instance, in mathematics (cf. SCHOENFELD, 1983, p. 367). We shall only list a few simple dimensions that could have played a role in our experiments on the 'base—rate fallacy'. These are, for example:

— Subjective time restrictions due to the subject's tradeoffs when receiving a certain reimbursement.
— The operation target of generating an algebraic solution (that has at least the appearance of precision).
— The goal of producing plausible arguments and justifications for each step.
— The goal of rating the reliability or credibility of knowledge or certain operations.
— The goal of not losing 'too much' precision through the estimatory nature

of rough calculations.
- The desired exactness of predictions (anticipatory precision). Such goals may be modeled with discrete aspiration grids (cf. SAUERMANN & SELTEN, 1962; SCHOLZ, 1980a; TIETZ, 1983).
- The desire for an aesthetically more attractive and more elegant solution (this involves comparative goals, cf. DÖRNER, 1976, by which the aim is to achieve the elegance of another known procedure).

Examples/Comments
a) As may be seen from this description of the Goal System, it is considered to be a reservoir of ordered goals. Within the presented framework no operations are performed within the Goal System. In a computer terminology it is regarded as a store. Clearly other approaches are also possible, but there are a variety of advantages in keeping the number of processors as small as possible if, for instance, a computer simulation is desired. Within the presented framework, the following assumptions may be formulated: Individual and situational differences in the actualized Goal System (e.g., an aspiration grid) may be modeled by different **structures** of this system itself, by different **preinitializations** of certain dimensions, and different tendencies to elicit certain goals. A possible precision and possible way of modeling these differences can be achieved by constructing a mathematical model of goal setting or goal formation behavior. If one proceeds analogously to the spreading activation theory formulated within ANDERSON's, 1983, "Architecture of Cognition", one first has to introduce a suitable structure onto the elements of the Goal System, then second, one may introduce a (random) function, g_1, a kind of search function, that maps from the Working Memory's present state into the Goal System (which initializes a field in the Goal System), and a (random) function, g_2, which models the loading of the Working Memory with goals.

(6) The **Evaluative Structure** (ES): This contains the operators, procedures, and heuristics used for the evaluation of a variety of activities and objects present in the Working Memory. There are different entities which may be the target of an evaluation. For instance, in a problem solving activity, the density of knowledge retrieval, the heuristics at the Working Memory's disposal, the goals or subgoals, or the level of attention or activation may be the object of an evaluation. We want to stress at this point that the elements of the Evaluative Structure operate in the Working Memory on specific entities with respect to specific goals which are currently actualized. Thus the object of operations performed by elements of the Evaluative Structure is by no means the whole Working Memory's activity.

An important activity of the Working Memory, which is governed by the

CHAPTER 5.1.1.

Heuristic Structure through the activation of certain evaluative functions, is the rating of the **credibility** (cf. HUBER, 1982, p. 137) of information which is present (via encoding or retrieval) in the Working Memory. The credibility rating has been a straight interpretation of the information weights in the individualized normative solution which has been presented as a model for a probability judgment in the base−rate problems in Chapter 2. As mentioned above, we consider goals under consideration to be the major basis for the elicitation of evaluators. This has been introduced because the dimensionality of the goals may be essential for the evaluative operators. The Evaluative Structure has not yet been extensively investigated, as HUSSY, 1983, notes. But there is some evidence within our experiments that evaluative functions are often not appropriately developed for the activity of probability judgments in base−rate problems. The long list of idiosyncratic algebraic operations which have been identified in Experiment A, for instance, may indicate that not only the individual's Knowledge Base and Heuristic Structure, but also the Evaluative Structure do not meet the requirements of the task in stochastic thinking.

It is conceivable that the Evaluative Structure is a very important element from a developmental psychology aspect, and that, for example, the leap from the so−called concrete−operational to the formal−operational phase (cf. PIAGET, 1972), is reflected in a qualitative restructuring of this structure. "It is only the ability to reflect on our own thinking and to recognize the insufficiency of our intuitive judgments that enables us to generate alternative and better strategies for handling problems." (SCHOLZ & WALLER, 1983, p. 307). The following illustrations feature genetic components:

Examples/Comments

a) This illustration concerns probability judgment and is of a more historical nature. The story of the Chevalier de Mere is an anecdote that is found in many basic probability text books (cf. e.g., FREUDENTHAL, 1973, p. 528). The Chevalier, a passionate gambler, was concerned with the question as to whether it is advantageous to bet on the occurence of at least one double six, within 24 throws, using two dice. Apparently, it was known during his time that it is advantageous to bet on the occurence of one six within 4 throws, using one dice. Presumably, the following false heuristic was applied: If one dice is thrown 24 times, one could expect an 'average' of 4 sixes. If a second dice is also thrown, the chance of it being a six, during one of the average 4 events when the first dice should give a six, is at least 0.5, as it is advantageous to throw a six in 4 trials. In actual fact, the probability of throwing a six within 4 throws using one dice is 0.518, while the probability of throwing a double six, within 24 throws, using two dice is 0.491, which is only very slightly lower than .5.

One version of this anecdote records that the Chevalier's heavy losses were the cause for the investigation of this problem. It is doubtful whether such a goal – expectation – state discrepancy brought up the problem and the consequent correspondence between de Mere and Pascal, as the differences between the probabilities are very small and the variances are rather large. A second more plausible version is that the Chevalier's heuristic was confronted with contradictions or a subjective fuzziness in the justification, and that this brought about the investigation of the problem and a clear proof via considering, for instance, the binominal distribution.

b) Like the Goal System, the Evaluative Structure is modeled as a reservoir of operators and not as a processing unit. Remarks similar to those in Comment a) on the Goal System apply here.

c) We distinguish between two types of operators within the long – term stores, namely elements of the Heuristic and of the Evaluative Structures. Hence one has to ask which criteria separate these two structures. As mentioned above, we regard the stores to be functional units. The functional dichotomy between these two operators is established by the distinction between progressive and evaluative activities. When coping with a problem, "a problem solver is in a progressive phase when he or she works directly toward the goal state of the problem. By contrast, the problem solver is in an evaluative phase when he or she evaluates some already performed part of the problem solution" (ALLWOOD, 1984, p. 414), or if he or she evaluates the usefulness of **heuristics** or the **credibility** of information. Hence the objects of **evaluative operators** are results of the inferential process (with respect to specific goals), knowledge elements (with respect to their credibility and usefulness), or available heuristics (with respect to their usefulness). **Heuristics** themselves usually operate on the information given and the knowledge retrieved. We wish to note that we consider the assumption of completely separated Heuristic and Evaluative Structures would be an overstressing of our multistore conception. Our approach emphasizes the functional aspect of the use and elicitation of operators. Presumably an adequate representation would consist in an understanding of the Evaluative Structure as a particular dimension of an amended Heuristic Structure perhaps similar (though on another level) to the distinction between simple and higher ordered heuristics. Nevertheless, it is hypothesized that the Heuristic and Evaluative Structures essentially differ from a functional, acquisitional, and also genetical (as doing preceeds evaluation) point of view, and thus a conceptual separation of heuristics and evaluators will be introduced.

(7) The **Guiding System or Central Processor** (CP): Within the presented

framework the Guiding System governs the activation and the interaction of the individual elements of the information processing system. In the tradition of the computer analogy, HUSSY, 1983, p. 51, called this unit the Central Processor and described its functions as follows:

"Through its governing and controlling functions, the central processor embodies the process element of information processing" and contains "complex heuristic and algorithmic processing strategies" (our translation). What form these (algorithmic) processing strategies take, is however not specified by HUSSY. Within this text we will predominantly deal with the Central Processor's **quantitative** and **qualitative** governing of the activation of other cognitive units. Such a governing system could function in two directions. From one direction (bottom−up), the state or the concrete work situation in the Working Memory influences this governing. For example, a breaking off of intensive memorization phases could be allowed in order to undertake more intensive external information search if one (i.e., the individual's Working Memory) cannot get any further with the information already available (cf. Chapter 4.5.3.). From the other direction (top−down), a change in the interplay can also be brought about through the Guiding System by general cognitions about the situation under consideration (e.g., by categorizing the problem type). In the previous chapters (cf. Chapter 3.2.), we have introduced the analytic and the intuitive modes of thought. A switch between these modes of thought may be conceived as such a change, and the above remarks may be interpreted in the sense that an alteration in these states or modes is on no account based only on the data (i.e., induced by the direct input stimuli) but is also brought about by the internal reconstructions of such stimuli (i.e., the specific relationships between task, environment, and individual can lead to changes in the entire activity). Such a breaking off then leads to a changed preinitialization or predisposition that is initiated by the Central Processor.

As mentioned above, within a conceptualization of the Guiding System in a computer analogy, one has to specify which "complex heuristic and algorithmic processing strategies" are actually present in the Guiding System. Similarly, within our system theory approach of knowledge−based information processing, we have to point out the basis on which the preinitialization and activation shifts are determined by the Guiding System, and how these shifts are to be conceived. Naturally, we have to postulate knowledge and evaluators for judging both the Working Memory and the Decision Filter's activity. We have not mapped corresponding units for this knowledge and these evaluators in the framed flow chart (cf. Figure 5.1.), but consider them to be part of the Guiding System box.

In order to avoid misunderstandings, we wish to stress that the objects of processing and evaluation differ from the processes which take place in the Working Memory. The objects of processing are the Working Memory's (and the Decision Filter's) activity and information processing, whereas, for instance, evaluative processing in the Working Memory operates on specific entities within the Working Memory. Hence, processing in the Guiding System requires knowledge about information processing and its adequacy in respect to the person — environment relation.

The assumption of the existence of a Guiding System, a Central Processor, or an operating system, to introduce another term from the computer framework in our information processing model, also allows for a conceptualization of individual reflexivity and governing of states of consciousness and modes of thought. Basically, it seems that it is possible to model reflexivity and consciousness through an interplay of the Guiding System and the Working Memory; "consciousness can be explained through the recircuiting of other structures with the structure (which is to be made conscious)." (SEILER, 1984, p. 2). In principle, this can also be modeled in the structure presented in Fig. 5.1. (through the circuit of the Guiding System and the Working Memory). Thus we consider that it is useful to assume the existence of a Guiding System, as we particularly regard the governing of information processing, and thereby the active side of cognitive activity, to be very important within the frame of limited rationality theories (cf. SCHOLZ, 1981).

This approach is also in line with JOHNSON — LAIRD's, 1983, p. 503, "Computational Analysis of Consciousness". Within his computer analogy approach, he points out that, "a sensible design ... is to promote one processor, to monitor the operations of others ...", if the problem of control, self — awareness, self — reflection, or consciousness has to be modeled. At a first glance, this approach obviously may lead the problem of infinite regress, that means that a Super — Guiding System has to be postulated to monitor the Guiding System, and that a Super — Super — Guiding System has to be ... . However, as JOHNSON — LAIRD argues, the idea of recursiveness allows for a modeling of a "high — level processor" (i.e., the Guiding System) that controls the lower — level processors. These lower — level processors within our framework are the Working Memory, Sensory System, and possible also motor processors that are responsible for the individual's input — output behavior. We will not pursue these conceptual problems within this text but turn to the Examples/Comments.

Examples/Comments
a) The Guiding System is particularly involved in the cognitive framing of situations. This framing implies a preinitialization of the activation of the various cognitive systems.

In Experiment C, Hypothesis 1 on the impact of the Experimental Framing (cf. Chapter 4.5.3.) was based on the assumption that the preinformation about the nature of the situation (for instance, that one has to participate in a problem solving experiment or has recently worked on certain problems, cf. CHAIKLIN, 1985) induces preinitialization of certain domains or levels in the Knowledge Base, Heuristic Structure, Goal System, and Evaluative Structure. It is thus assumed that the Experimental Framing of stochastic problems very much focuses on a top–down activity. But the Content Framing (cf. Chapter 4.5.3.) is also supposed to produce a top–down framing, as the contents (e.g., hit parade or defectiveness rate of a motor) elicit a general disposition toward, and a 'global judgment' about, the adequacy of a certain problem treatment. Nevertheless, the initialization of goals in the Content Framing is assumed to function in a more bottom–up manner; that means, that the Working Memory's load specifies, for instance, the goals to be elicited.

(8) **Decision Filter (D), and Output, Action, or Overt Behavior (O):** In our model, we discriminate between the **decision** or resolution and the **Output** (cf. IRLE, 1975; SCHOLZ, 1980b). The decision or the intention (cf. FISHBEIN, 1967) to choose a specific alternative or to produce a specific reaction can be interpreted as being the last processing phase or the final product of the Working Memory that leads to an Overt Behavior. However, not every decision that is planned is actually carried out. Planned actions are sometimes not carried out because of emotional inhibitions such as fear, or simply because they are 'forgotten'. Because of this, we assume the hypothetical existence of a Decision Filter. This separation of action from action planning is a common assumption which can be found in cybernetic learning theory (cf. e.g., KLIX, 1971, p. 352; 1980), in social psychology theory (cf. IRLE, 1975, p. 32), and action research (cf. OESTERREICH, 1981). As, for example, IRLE, 1975, has pointed out, resolutions can also change internal personal states, the Evaluative Structure, values, motives, and attitudes, without producing any Output. Research into stochastic thinking needs to pay particular attention to the discrepancy between internal action, planning, and output, because of the only partially conditional nature of stochastic feedback and the difficulty in communicating uncertainty.

Examples/Comments
a) In order to illustrate the delicate problems involved in the discrepancy between action planning and action, we will include the feedback in this example. Let us assume that the alternatives A1 (don't act) and A2 (act), that are coupled with the subjective utilities u(A1) and u(A2), are repeatedly available. Let p1 and p2 be the objective probabilities for the two alternatives and s(p1) and s(p2) be the individual's subjective assessment of these probabilities. Now let s(p1)u(A1) < s(p2)u(A2), which may lead to the alternative A2 being chosen; but let p1u(A1) > p2u(A2), so that A1 is the better alternative.

If A2 is now repeatedly chosen, it will eventually lead to punishment for making a choice. If A2 is for specific reasons not chosen (for example, forgetfulness or motivational dispositions such as risk avoidance tendencies; cf. HECKHAUSEN, 1980), the neglect of one choice or action is rewarded and reinforced. The alternatives could be lotteries, insurances, and the like.

b) In many situations, particularly experiments, the freedom of action is restricted through the medium or the format of the action, or it must be transferred to a response format which is unfamiliar to the individual, or even does not permit the subject to respond in the way that he intends. In our experiments on the 'base–rate fallacy', the subjects were requested to formulate their answers in percentages. However, if individuals are only used to forming qualitatively ordinal probability judgments (cf. ZIMMER 1983; HUBER, 1983), then a translation is an essential process that can lead to considerable misinterpretations (BEYTH–MAROM, 1982) and distortions, as the subject may eventually fail to attribute any meaning to a probability in percentages. Within the experiments reported in Part I of this volume, the response format of a percentage judgment was nevertheless chosen, as this format is, in the author's opinion, presumably the most popular one for the sample of students investigated. We suppose, however, that in probability judgments an obligatory response format may be the source of inhibitions by the Decision Filter.

c) The introduced conception of the Decision Filter only refers to decisions which intentionally result in Overt Behavior. Of course, intermediate products of the inferential process or judgments which are only used for 'inner processes' may also be suppressed or filtered. As outlined in the section on the Evaluative Structure, however, these processes are modeled as part of the Working Memory's activity through the application of evaluative procedures.

Concluding remarks on the framework's units. As outlined in this section, the nine units which have been introduced have different status. The main **processing units** of the proposed framework are the Working Memory (including a Short–Term Memory) and the Guiding System.

The four units of **Long–Term Storage** (LTS) are the Knowledge Base, the Heuristic Structure, the Goal System, and the Evaluative Structure. These four units, which are framed by broken lines within Figure 5.1., may be ordered along two dimensions. First, the Heuristic and Evaluative Structures are assumed to consist of operators which may be applied to entities of the Knowledge Base, the Goal System, and of course encoded information. Second, we distinguish stores used in the direct coping with the problem under consideration (i.e., the Knowledge Base and the Heuristic Structure) from stores which are involved in the evaluation and direction of the cognitive system (i.e., the Evaluative Structure and the Goal System). There are various reasons for these distinctions: For instance, the acquisition of new operations takes more time than the learning of new concepts or new goals. Further, it is a common feature in clinical psychology that the Knowledge Base and the Heuristic Structure are extremely highly developed, while severe deficiencies may be identified in goal setting behavior and self–evaluation (cf. BRATUS, 1978). In general, we presume that in both dimensions a different length of time is required for the acquisition of new elements. On the one hand, new knowledge elements and new goals are usually acquired more quickly than heuristics or evaluative operators. On the other hand, goals and evaluators are acquired more slowly than the corresponding operative entities (i.e., elements of the Knowledge Base or Heuristic Structure).

Finally within our cognitive framework there remain the Sensory System, the Decision Filter, and the (overt) Action, which may be regarded as both part of the individual and his or her environment.

5.1.2. THE CONNECTIONS BETWEEN THE INDIVIDUAL COMPONENTS

The bracketed numbers in the following refer to the relations mapped as pathways in the framed flow chart (Figure 5.1.).

Internal relations: The informations that are received by the Sensory System are filtered and encoded. A selection from these informations is temporarily stored in the Short–Term Memory which is part of the Working Memory. During the activation of the Sensory System, the Central Processor is governing the level of activation (2), and, if necessary, also governing the activity level of the Working Memory (3). Under specific conditions, this can lead to a selective perception, even during the very first stages of information processing.

Knowledge retrieval depends on the specific Working Memory load and the preinitialization of the Knowledge Base (see above). The relation of knowledge retrieval has two sides, namely the activation of elements or domains by the Working Memory (4), and the feeding in or firing of knowledge elements from the Knowledge Base (5). Obviously, the process of knowledge retrieval may be a process of planned knowledge search as well as an unguided retrieval process.

If a **problem** occurs in the Working Memory (i.e., if the questions present in the Working Memory cannot be answered by a direct consultation of the Knowledge Base), and an activation search process for heuristics commences (7), certain heuristics which may operate on the Working Memory load are fed to the Working Memory (6).

The intermediate results obtained in the Working Memory are evaluated. This is considered to be done in the Working Memory when applying evaluative operators which have been fed to the Working Memory (10) by the Evaluative Structure. As mentioned above, various entities may be elements of the evaluative operators' variable space. Clearly, the evaluation of internal or external states needs a point of reference. These points of reference are provided (8) by the individual's Goal System.

Under certain circumstances (cf. the chapters about the modes of thought), the evaluation procedure in the Working Memory results in a search for new goals (11). This is particularly the case if intermediate results or states of the Working Memory are judged to be unsatisfactory. As we know from various fields of psychology, goal setting behavior is by no means an arbitrary procedure (see the section on the Goal System above). Some conclusions may be derived about both the structure of the Goal System and the Working Memory's search and elicitation process (9) in the Goal System.

If a product of Working Memory is evaluated as being compatible with the goals under consideration, it is uncoupled as a decision, transferred to the Decision Filter (12), and the decision is eventually stored in the Knowledge Base (5). The Central Processor continually receives information about the activities of the Working Memory (13). We further postulate that the Working Memory and the Central Processor receive feedback over the decision made by the Decision Filter (14) and (15). Based on this information, the Central Processor induces initializations in the Knowledge Base (16), the Heuristic Structure (17), the Goal System (18), the Evaluative Structure (19), and the Decision Filter (20), dependent on the state of the Working Memory and the individual history of decision making. In our opinion, this initialization by the Central Processor may be regarded as a global governing, while the initialization processes of the Working Memory are of a specific and local nature [cf. Comment b) on the Goal System and Comment a) on the Guiding System]. The governing processes of the Central Processor may be regarded as assimilation procedures of information processing and cognitive recognition. As mentioned repeatedly above, they are not only undertaken in a bottom—up quasi—data—dependent manner (via 4, 5), but are also undertaken through reflexivity; the individual's self—determination with regard to the relationship between himself and his environment. Within our model, this reflexivity is mainly modeled by the Working Memory — Central Processor loop. This loop establishes a second or higher order relationship between the individual and the environment. We will call this relationship between the individual and his environment cognitive framing and identify it as the external relation A2 in our structure—process model of human information processing. There are in addition two further external relations A3 and A4. A3 represents the relation between cognitive decisions and the (overt) Output, an act of speech, sensory activity, or a motor action. As this Output can be regarded both as an Action and as a component of the individual, it can be viewed as standing exactly on the border between the individual and the environment. Actions themselves lead to changes in the environment which, along with other environment stimuli, can be perceived as feedback, A4, in our model.

5.1.3. SOME REMARKS ON THE FRAMEWORK'S RELATIONSHIP TO OTHER APPROACHES TO A MODELING OF INFORMATION PROCESSING

Clearly our framework is in line with the information processing approaches as developed by NEWELL and SIMON, 1972 (for an overview compare LINDSAY & NORMAN, 1977; or SCHAEFER, 1985). When discussing the specific relationships between the proposed framework and other approaches, HUSSY's SPIV—model (cf. HUSSY, 1983, 1984) must receive first mention. Like our framework, the SPIV—model postulates two hierarchically connected processors, a Working Memory and a Central Processor, and several epistemic structures which constitute the long term—knowledge and are structurally connected in a similar way to our outline. The main differences between the SPIV—model of general information processing and our

framework for the process and structure of information processing, which has been designed in order to generate an approach to an understanding of the specificities of stochastic thinking, are as follows:
a) the supplementary proposition of a Goal Structure,
b) the supplementary proposition of a Decision Filter,
c) the introduction of different levels of knowledge and heuristics, and
d) a different conceptualization of the Guiding System's functions (cf. the section on the Guiding System above) including cognitive framing which has been introduced as a crucial activity of the Guiding System in stochastic thinking.

Point d) is considered to be an essential difference. Within our framework, most of the information processing, for instance, the concrete problem solving or the generation of probability judgments, takes place in the Working Memory. Processing in this unit is conceptualized as an interplay of knowledge elements and goals with heuristics and evaluative operators. The latter operate on the former if they are applicable for the specific Working Memory load. Thus our approach differs from HUSSY's and others who assign an overall guiding and monitoring activity (for the input and operation of elements from the Long−Term Stores) to the Central Processor and hence, in our opinion, overburden this unit.

A similarity, that is comparatively weaker than that with the SPIV−model, may be seen in the work of DÖRNER, 1979, or ANDERSON, 1983, although these conceptions have to be assigned to different branches of cognitive science (namely the psychology of problem solving and artificial intelligence) and also feature a different level of concreteness. However, a multistore conception which is salient for their approaches is a common feature with our framework. The need for an introduction of a separate Goal System, which is also tied to motives and emotions, within a modeling of higher cognitive processes has been pointed out by KLIX, 1980.

5.1.4. SOME REMARKS ON THE FRAME CONCEPT AND ON COGNITIVE AND ON SITUATIONAL FRAMING

As mentioned above, the concept of framing receives different treatments within this volume. When the concept of Problem Framing was introduced in Chapter 4.2. (cf. p. 107) some remarks on framing, in particular on its use in problem solving and decision research, were also given. In order to illustrate the common denominator of the various approaches to defining framing and related concepts, we want to add some brief notes which may help in understanding our use of this term, for instance, when used to explain the activation of and a switch between different modes of thought.

MINSKY's (1975) concept of framing is similar to the concept of schema, but more specifically refers to a subset of the phenomena encompassed by the term schema. It emphasizes connectedness as a salient and important part of experience. Frames incorporate remembered knowledge and stereotyped situations. The concept of **framing** as a **situational** variable, as applied in Chapter 4., refers to these stereotyped situations.

MINSKY, 1975, p. 212, originally defined a **frame** as "a data−structure for representing a stereotyped situation" wherein a data−structure is conceived as an **internal** cognitive structure. Thus, in some respects, the frame concept is essentially identical with other terms such as "schema" (cf. BARTLETT, 1932), "script" (cf. ANDERSON, 1983), "gestalt" (cf. WERTHEIMER, 1945), etc. The differences between these terms only involve the areas, forms, and/or contents to which they are applied. For example, SEILER (1984, p. 92) points out that the older term "gestalt" describes the structure of relations in a more static isolated form, whereas, for instance, the term "plan" additionally emphasizes the dynamic aspects. The wide ranging term "schema" is used in different ways; for example, it is used to describe both perception and action schemata (AEBLI, 1980, 1981). However, the above−mentioned terms have one thing in common; they describe internal organizational structure and order without which cognitive recognition and action would be inconceivable, and they describe the iteration and 'new−actualization' of such structures (schemata) that belong to the basic forms of cognitive activity.

One can find a top−down and a bottom−up, a backward and a forward aspect in the frame concept. On the one hand, framing is essential for the understanding of observed sequences of events, while on the other hand, it is used to guide behavior. Analogously to ABELSON's (1978; quoted in CARVER & SCHEIER, 1981, p. 71) concept of a metascript, the cognitive **framing** of a problem or situation, as modeled in our framework, allows event clusters and episodes to be variable rather than constant, as it offers a frame system (cf. BARTLETT, 1932, p. 310; MINSKY, 1975; BROMME & STEINBRING, 1981) which is necessary to make information processing at all possible. One and the same fact, situation, or episode can be differently perceived, processed, and interpreted if it is differently framed.

5.2. MODES OF THOUGHT AS DIFFERENT ACTIVATIONS OF THE COGNITIVE SYSTEM

Originally, modes of thought in stochastic were introduced by lists of features. A clear deficit of this approach is that no structural relationship between these features has been defined. Although in two experiments (cf. Chapters 2.6.5. and 3.3.5.) cluster analyses consistently revealed a specific affinity to emotional reference, the genetical and procedural aspects of the modes of thought have been neglected. For instance, the definition via the lists of features did not allow for an understanding of the elicitation of, or of a switch between, the modes of thought.

There are two ways to redefine the modes of thought. We will start by reformulating the single features within the terminology and the structure (i.e., the components and the relations) of the framework. Then we will briefly sketch the modes of thought within the dynamics of the model, and elaborate the new perspective on the process of stochastic thinking and the results of the previous chapters. In order to provide a global impression of the structural differences between the two modes of thought, Figures 5.2. and 5.3. illustrate on which level the relations between the components are considered to be activated in the prototypical analytic and intuitive mode. We have chosen four degrees of relational strength to roughly represent the differences between the prototypical mode – specific activation levels in an ordinal manner (cross-hatched black lines, shaded, double open, and single lines, in descending order of strength). In order to avoid misunderstandings that might be raised by a first glance consideration, we want to note that the intuitive mode is not conceptualized as being just the same as an analytic mode but on a plainer and lower level, but rather as an essentially different activity in which other domains of long–term stores are involved. As pointed out in Chapters 3 and 4, for instance, no general behavioral superiority of one of the two modes is to be expected in stochastic thinking.

The following text presents a description of the single features in the left-hand column that may be regarded as redefinition of the lists presented in Chapter 3.2. (Table 3.1.). The two right–hand columns indicate the components and relations concerned.

CHAPTER 5.2.

The feature is established by:

	the components	the relations

A) Preconscious vs. conscious

Preconscious **processing** implies that there is only a weak self−reflection on the Working Memory's activity. This means that there is hardly any re−reference via the Working Memory−Guiding System loop. Evaluative local guiding is neglected during intuitive processing, hence also decisions are sometimes made with insufficient information and without any evaluation. The preconscious processing is also characterized by a weak and rather unsystematic setting of goals.

Preconscious information **acquisition** is presumably not accompanied by any intentional systematic preadjustment of the Sensory System and the Knowledge Base (with regard to the internal retrieval of information) during the course of information encoding.

Oppositely, guided information search is a feature of a conscious information processing. This may result in an action planning in order to provide systematic new information or in the use of external sources or stores (e.g., using paper and pencil). Conscious information processing is a guided, controlled, and goal−directed activity which is also continuously evaluated in the Working Memory. However, conscious information processing also often means that the Working Memory's activity is evaluated and adjusted by the Guiding System. This is also modeled by a recircuit of the Working Memory via the

components	relations
W − CP	3,12
ES	10,14
GS	8
S	2
(KB)	(16)
CP − S	2
GS,ES	8,9,10,11

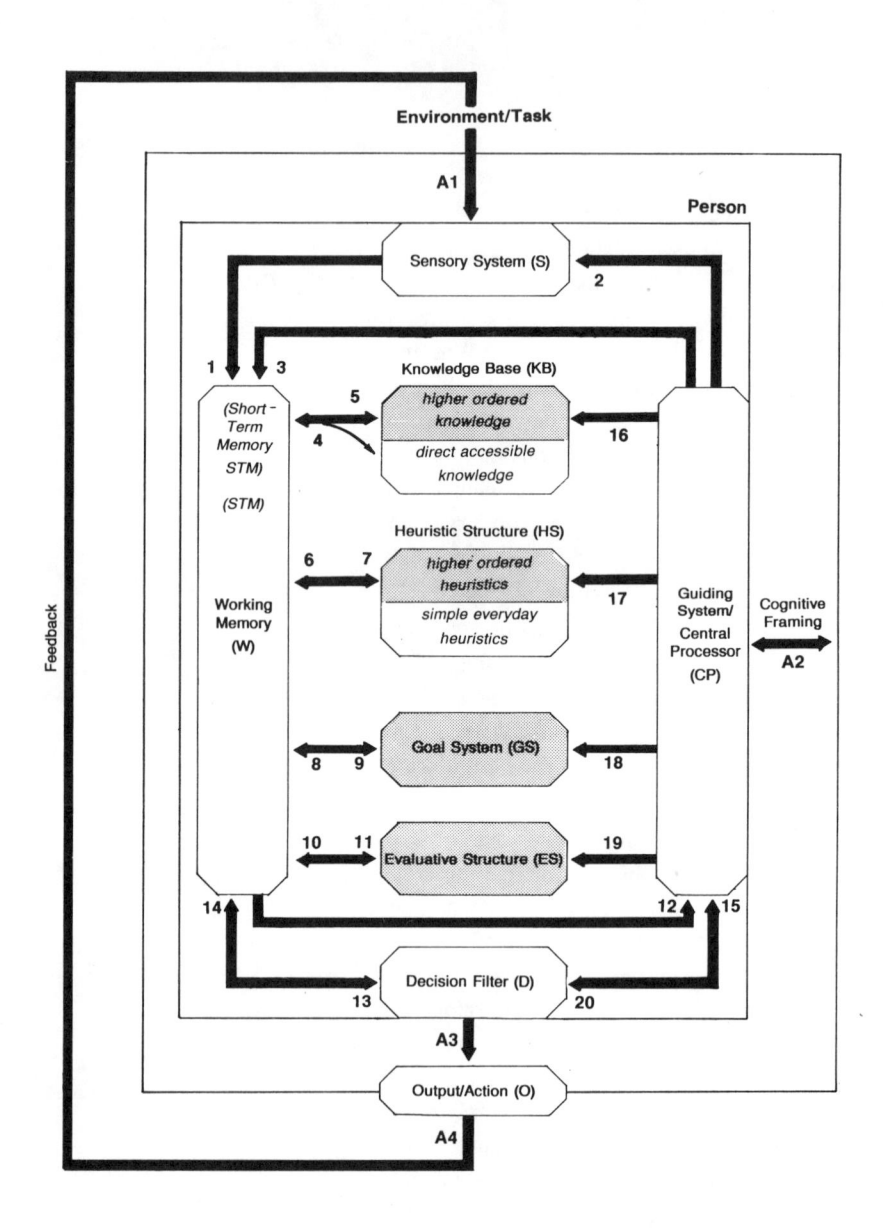

Figure 5.2.: The hypothesized activity and the intensity level of the interplay between the units in the **analytic mode** of thought

CHAPTER 5.2.   173

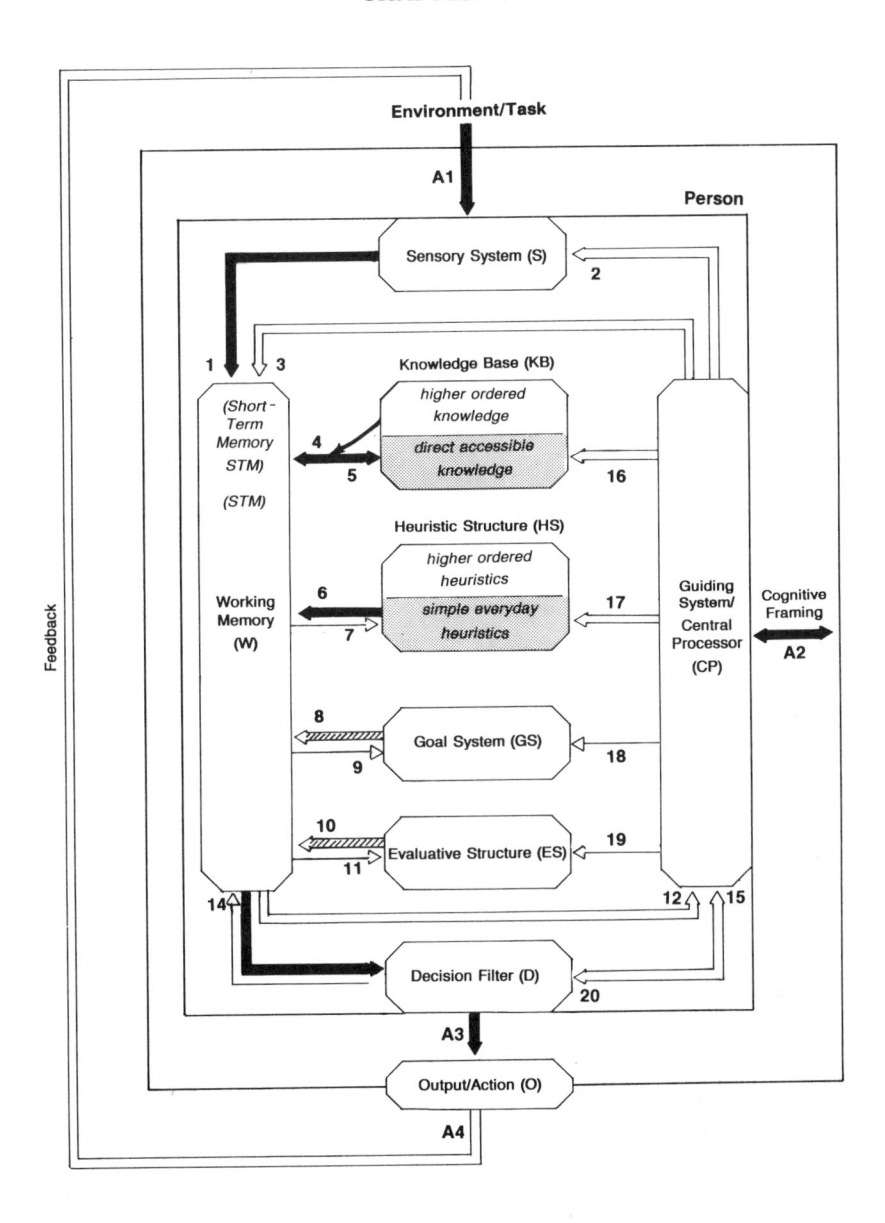

Figure 5.3.: The hypothesized activity and the intensity level of the interplay between the units in the **intuitive mode** of thought

Working Memory — Guiding System loop. Furthermore, we assume that information about the decision or output filtering is fed back to the Working Memory.

B) Understanding by feeling vs. pure intellect

Intuitive information processing is supposed to be attended by emotional goals, e.g., that alternatives are chosen which lead to a better feeling or which give the impression of being alright. If the latter is expressed in the framework's terminology, one would say that the activated set of evaluative operators is specified, and that emotional evaluators are operating in the Working Memory. The wording "understanding by feeling" implies, however, not only the involvement of emotions, but also a certain penetration into the problem.

As we know from mathematicians (cf. POINCARE, 1929; FISCHBEIN, 1982), this particular feeling is no superficial process, but has to be accompanied by an extended dealing with the problem structure on hand, or previous coping with similar problems, that may provide the basis for a subjective feeling of having grasped the problem.

The dichotomy to the analytic pole is predominantly established in this feature by heuristics and evaluative operators which may be denoted as being logical and purely intellectual. By intellectual operators we mean operators which are explicitly defined and which permit a codification that is independent of the individual's feelings and emotions.

C) Sudden, synthetical, parallel processing of

D	14
GS	8
ES	10
HS,ES	6,8

a global field of knowledge vs. sequential, linear, step−by−step ordered cognitive activity

This dimension focuses on the knowledge retrieval and the heuristics that are used. Within the intuitive mode, there is no pre-initialization on a certain domain of the Knowledge Base, and peripheral elements of the Knowledge Base (usually elements of lower ordered knowledge which are only marginally related to the actualized knowledge) are also taken into consideration. In contrast, a systematic, local, 'neighbored' information search is typical for analyticity.

K   4,5,16

The types of heuristics also differ: simple, everyday, fast processing heuristics using the redundancy of information or rules of thumb (like rough middling strategies) result in sudden **intuitive** decisions, whereas sequential, linear, usually high capacity requiring heuristics are involved in **analytic** thinking. The higher ordered heuristics are assumed to be activated in the analytic mode.

HS   6,7,17

D) **Treating the problem structure as a whole vs. separating details**

If the 'Gestalt' is recognized, the whole situation is considered and no single information elements are separated when working on a problem or task. Clearly, some conditions have to be fulfilled if a 'Gestalt' is to be adequately recognized. In this case the problem has to be within the individual's ability. This means, in particular, that the Knowledge Base has to meet the requirements of the situation, and a multitude of cues in the Working Memory mirror and restructure the situation.

W−KB   4,5

The separation of elements may be both the product and the basis (or starting point) for analytic heuristics.

E) **Dependent vs. independent of personal experience**

The personal idiosyncratic interpretation or problem representation of a problem or situation in the Working Memory is typical for the intuitive mode. Such a representation is provided by personal representatives, that means for instance, that if the problem concerns an automobile, the subject's own car is imagined. In general, the Working Memory is reloaded with knowledge elements or episodes (in the sense of TULVING, 1983) which have connotations of personal experience. Such a reloading may even start during the very beginning of information encoding.

The dimension of personal experience also arises in the actualized Evaluative Structure and especially in the goal or aspiration formation which is highly self–centered in the intuitive mode, and usually of an abstract, formal, impersonal character in the analytic mode. In particular, the 'dimensionality' of both the Evaluative Structure and the goals is assumed to be dependent on the specific preinitialization through the Central Processor.

F) **Pictorial metaphors vs. conceptual or numerical patterns**

This above dichotomy may show up both in the kind of encoding and representation of knowledge in the Short–Term Memory, and the retrieval and representation of knowledge in the Working Memory. Without relying on a particular model of the Knowledge Base,

HS	6,5	
KB	4,5	
S–W–KB	1,3,4	
G,ES,CP	6,17,8,18	
W,KB	4,6	

we propose that pictorial, visual, concrete, and episodic information or knowledge elements are characteristic for the intuitive mode, and abstract, conceptual, verbally stored information (the so—called higher ordered knowledge) for the analytic mode. Of course, the activation of this dichotomy is assumed to be a function that is controlled by the Central Processor.

CP   16,(3,4)

G) **Low vs. high cognitive control**

The intuitive mode is usually a rather weakly controlled activity. In particular, the local control of the Working Memory's activity by the Evaluative Structure, which is typical for the analytic mode, scarcely appears during intuitive processing. Further, the intuitive mode is normally either an (positively) emotionally specified (see below) or a relaxed activity. The latter is often characterized by a minimum of drive and goals.

ES,CP   10,11,19

GS   8

In the analytic mode, it is assumed that the Decision Filter is feeding its activity to the Central Processor, and the results from this unit's processing are also fed back to the Working Memory. Therefore, only well—considered decisions pass this filter and result in actions. High control is also reflected in a high activity level in the Central Processor which leads to the activation and initialization of all the other units.

D—CP   14,15

CP   3,16,17, 18,19,20

H) **Emotional involvement without anxiety vs. cold, emotion—free activity**

A shift toward an emotional involvement produces complex changes in the information processing. According to recent literature on the emotions (cf. e.g., KUHL, 1983a, b, obviously relying on IZARD, 1977;

CP

FIEDLER, 1985), the most crucial changes brought about by an induction of emotions have to be expected in the Central Processor, and hence indirectly in the complete system, yet predominantely in the Evaluative System, the Goal System, and the Decision Filter. A good description of the impact of emotional involvement on information processing is given by BASTICK, 1982, p. 117. He proposes, within his theory of emotional sets, "Judgments of similarity or suitability are made by the relative ease by which one may go from being in one emotional set to being in another, a subjective evaluation of the transition probability of going to emotional set B given being in emotional set A." ES,GS

Further, it is known that in many decision activities a "positive feeling" and the lack of fear, stress, and anxiety often results in unjustifiably high aspirations (cf. SCHOLZ, 1980a) and goals and an overrating of one's resources. HS,GS

On the contrary, according to BASTICK 1982, a "cold" processing activates analytic systematic heuristics and evaluations, as well as a systematic search in special domains of the higher levels of the Knowledge Base through focusing or clustering. Which differential emotions, or whether any at all are tied to analytic thinking is still a disputed question which will be discussed in the following pages.

However, first we will turn to a brief **description of the modes of thought within the dynamics of the framework**

Each of the modes of thought's features specifies the quality and intensity of relationships within the information processing system (cf. Figures 5.2. and

5.3.). We believe the **intuitive mode of thought** to be the way of thinking that is most naturally chosen unless an analytic mode of thought is induced by a switch through the Guiding System. In the intuitive mode, information is sampled in a relatively unguided haphazard manner. Within the Knowledge Base, the direct accessible knowledge is retrieved and used in the encoding process. There is no systematic search in either the Knowledge Base or the Heuristic Structure, and mostly simple, fast processing, everyday heuristics are applied. The Emotional System, which is in a "positive state", is considerably involved in the goal formation. The Evaluative Structure, the setting of the Decision Filter (low control), and also the guiding activity of the Central Processor (or the Guiding System, which preinitializes on the 'natural domains' of the Knowledge Base and the Heuristic Structure) are not completely bypassed, but are only integrated on a very low level.

As is shown by Figure 5.2., **analytic thinking** is founded on an activation of the cognitive system in which more units and relations are involved. The domains of higher ordered knowledge and heuristics are preinitialized (top—down), and the Evaluative Structure is activated so that evaluative operators are fed with high density to the Working Memory. The information acquisition itself is accompanied by a guided selective information search and a systematic focusing exploration in the Knowledge Base (bottom—up) within the preinitialized domain. Sequential, higher ordered, and sometimes even formal heuristics are made available, and the information processing is accompanied by a self—reflective, self—controlling monitoring and supervision. The decisions are considered, controlled by the Decision Filter, and fed back to the Working Memory and the Central Processor.

On the definition of cognitive strategies. The modes of thought may be considered as a main dividing line within the set of cognitive strategies that are applied in stochastic thinking. Withing this text, the concept of cognitive strategy has been implicitly defined by the various modeling approaches, heuristics, and case examples that are reported in the previous chapters. However, the introduced framework permits a more explicit definition. When referring to the game theoretical strategy concept (for games in extensive form, cf. OWEN, 1972), we may define a cognitive strategy as a complete decision plan which generates a decision through the use of specific knowledge elements, heuristics, goals, and evaluative operators for each state and stage of problem coping within the cognitive system.

On the role of emotions in the assessment of the modes of thought. Within the original definition of modes of thought by lists of features, a rough relationship between emotions and the activated mode was described. According to Feature H, an emotional involvement which excludes anxiety and fear is an attribute of intuitive thinking, whereas emotions signaling danger are assigned to the analytic side. But also some other features, for instance, Feature B (understanding by feeling vs. pure intellect) and, in some respects, Feature I (feeling of certainty vs. uncertainty) involve emotional aspects. We cancelled the latter feature from the lists of the modes of thought in stochastic thinking, as, in the experiments on the 'base−rate fallacy', this feature did not show any relationship to the other features on the intuitiveness/analyticity scale. We want to discuss, exemplarily for the aspect of emotions, what progress may be gained from the introduction of the above framework of the structure and the process of the cognitive activity that will increase our understanding of the significance of emotions in both modes of thought.

In our opinion, the proposed framework of the process and the structure of information processing allows a more precise theoretical conceptualization of the potential switching function of emotions in respect to the modes of thought. However, there are alternative theories on how the switching function might be conceptualized and which emotions trigger which mode. KUHL, 1983b, p.235, considers so−called primary emotions to be responsible for carrying out switching functions. When referring to IZARD's, 1977, taxonomy of emotions, he hypothesized that anxiety, fear, shame, guilt, and eventually surprise/startle are 'favoring' the sequential analytic mode, whereas 'positive emotions' like interest and joy should induce the intuitive mode. KUHL's conclusions are primarily based on phylogenetic arguments and secondary findings from experimental psychological research. Like ours, KUHL's framework is a system−theoretical one. Analogously to CRAIK and LOCKHART's, 1972, sequential level of processing model, he assumes four 'levels of cognitive activity': the analysis of physical stimuli, the elicitation of schemata, the analysis of conceptual−semantic features, and a propositional analysis. Each level of analysis is assumed to affect the emotional state, which itself influences and possibly alters the cognitive activity on the subsequent level. Although describing an interaction between cognition and emotion, and its presence through all phases of coping with a problem or task, no detailed insight may be gained from this approach into which of our model's units or relations are crucial for the switching process, and how these units and relations are structured.

A somewhat different and more differentiated view on the impact of emotions has been formulated by CLARK, 1982. Within her paper, she particularly stresses the impact of autonomic arousal. She emphasizes the difficulty of distinguishing positive moods from negative ones, for instance, discriminating joy from sadness by measures of blood pressure etc. In line with SCHACHTER and SINGER's, 1962 (cf. also SCHACHTER, 1964), assumption of emotion as an autonomic arousal plus a cognitive label, she proposes a model of arousal—induced automatic priming which elicits a process of spreading activation (cf. CLARK & ISEN, 1982), such as that described by COLLINS and QUILLIAN, 1969, COLLINS and LOFTUS, 1975 (cf. CLARK, 1982, p. 266). Knowledge is regarded as being emotionally toned and primed, and arousal itself is considered to be a cue to and of affective material. Within our model of information processing, CLARK's assumption would imply that arousal is stored in the Knowledge Base quasi as an object property. As CLARK, 1982, p. 279, emphasizes, the arousal is mostly unconscious, "without awareness, and without interfering with other ongoing processes" that are involved in judgment and behavior.

She also concludes that, "sometimes arousal may influence judgments and behavior through more effortful controlled strategies" (p. 282), and hence, emotions may be linked to the analytic mode. She introduces two examples of this. The first is the subject's need for an explanation of an arousal that has no obvious cause (cf. SCHACHTER & SINGER, 1962). The second is a subject's attempts to alleviate the discomfort of arousal. Hence again, the idea of an interaction of emotions and cognition becomes evident (cf. MEYER, 1983). While the hypothesized relationship between emotions and cognition may be clearly conceptualized within our model of the process and structure of information processing as an induction of systematic search processes in the Knowledge Base, etc., the latter is not the case for unconscious arousal, as the way in which awareness of arousal has been gained is not specified.

Based on our theoretical considerations, and also on an interpretation of the experimental findings (especially the effect of the framing variable), the following conceptualization of the interaction between emotional involvement may be hypothesized. The applied definition of different emotions refers to TOMKINS, 1962, 1963, and IZARD's, 1977, 1981, taxonomies of emotions (RUSSELL's, 1980, circumplex model of affects, with pleasure, excitement, and relaxation as positive; and displeasure, distress, and eventually depression as negative feelings might also be applied). A positive emotional state, for instance, the induction of joy and perhaps interest (cf. KUHL, 1983), is regarded as being responsible for a reduction of the control and Decision Filter activity and the application of emotional evaluators and goals. For

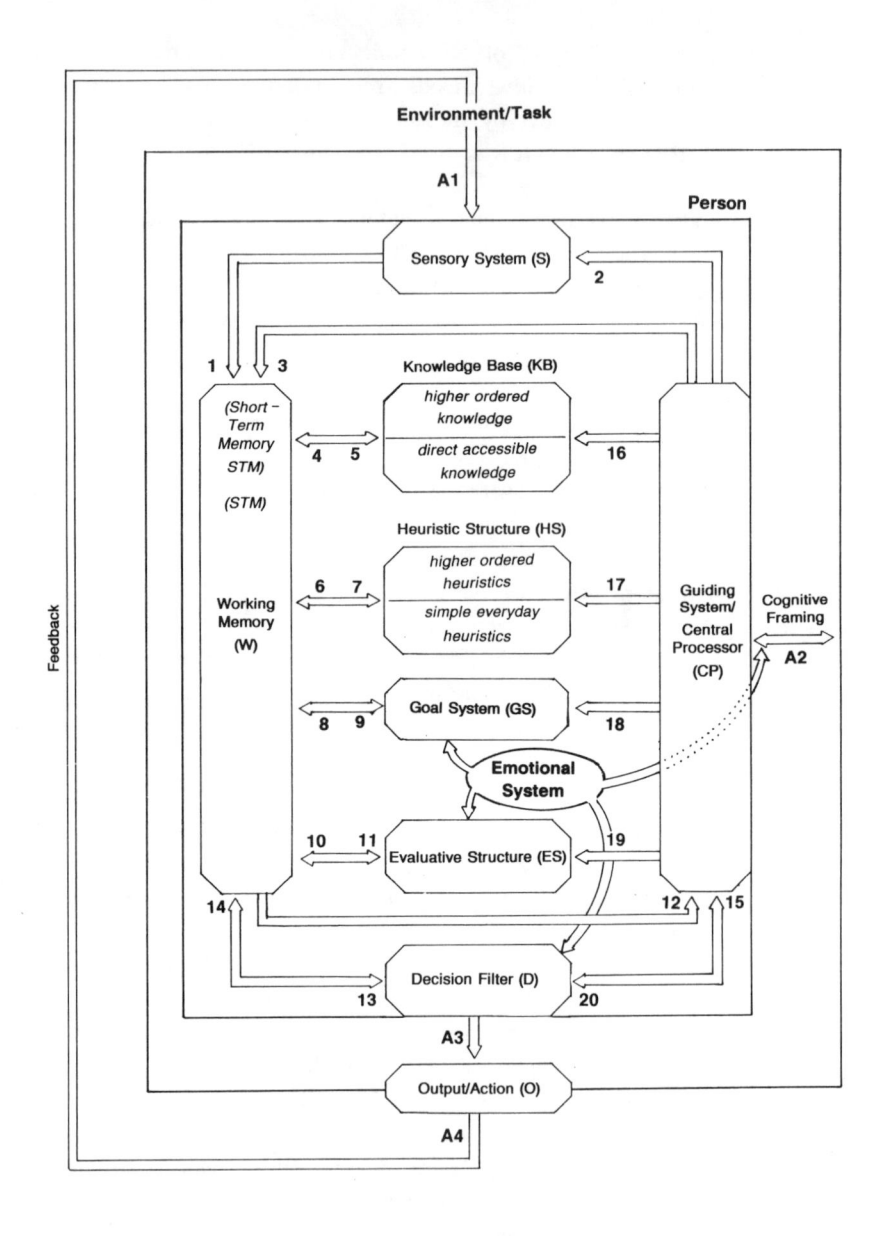

Figure 5.4.: Schema of the hypothesized interaction between emotional involvement and the cognitive system

instance, risky or dangerous alternatives are chosen in order to retain the current level of tension, excitement, or arousal necessary to accomplish an immediate success or satisfaction (cf. IZARD, 1981, p. 277, p. 292). But also negative emotions like anger or disgust are supposed to induce emotional goals and evaluators. These fundamental emotions are often accompanied by a reduction of control functions (cf. CARVER & SCHEIER, 1981, p. 178) and presumably also of decision filtering. On the other hand, emotions like sadness, distress, shame, guilt, and perhaps low or medium levels of fear and anxiety, might induce rational evaluators and goals and a systematic search for reasons that can explain certain events and thus produce a switch to the analytic mode.

Admittedly, the above considerations are of a rather theoretical nature and a controlled test of, for instance, changes in the Goal System by inducing alternating emotions will be confronted with the problem of treating emotions as independent variables (cf. ASENDORPF, 1984), although some researchers are optimistic that they may overcome this problem (cf. e.g., SCHWARTZ & WEINBERGER, 1980; SCHWARTZ, WEINBERGER, & SINGER, 1981). However, the path toward an experimental test of the above propositions and the impact of emotional involvement on stochastic thinking is straightforward. Figure 5.4. roughly illustrates our theoretical considerations on the main relationships between emotions and the forms of thinking in which stochastic thinking may appear. Clearly, our propositions do not exclude CLARK's assumptions about the facility in retrieving similarly toned knowledge from the Knowledge Base, but we regard different units to be critical for an emotional impact.

One may now speculate about the **confidence** feature which did not fit into the rating procedure (cf. Chapters 3.3.5. and 4.5.5.). Based on our conceptualization of the modes of thought, not only does confidence in the (analytic) methods or heuristics applied (cf. HAMMOND, HAMM, GRASSIA, & PEARSON, 1983) have to be distinguished from certainty with regard to the product of thinking, but also the two modes' ties to different emotions might be of essential importance. Emotions like happiness, joy, pride, or anger are supposed to be bound to the intuitive mode. Perhaps it might be plausible that a spontaneous act caused by extreme arousal or anger leads to an immediate satisficing (for instance in the case of an 'obviously justified punishment', which results in a high confidence in respect to the adequacy of the decision. Eventually, interest, as an emotion, might also be the source of insight and of the warm feeling of 'being right' long before rationality occurs (COURTNEY, 1968, quoted in BASTICK, 1982). Analytic thinking is a conscious process. An individual's feeling of certainty with regard to a procedure or the result of a decision or solution may also be the consequence of knowing the appropriateness of one's skills (i.e., Knowledge Base and Heuristic Structure). This process may also occur without a considerable emotional involvement but with an awareness of self—confidence and certainty. Obviously, a more differentiated theory about the feeling of certainty is needed, and the global question "How certain are you that your answer is correct?", which was applied in the experimental procedure of Experiment C and in many investigations into the individual's confidence (BASTICK, 1982), is not suitable to discriminate between the intuitive and analytic modes of thought.

5.3. THE NEW PERSPECTIVE AND ITS VALUE FOR AN UNDERSTANDING OF THE PROCESS OF STOCHASTIC THINKING

The conceptualization of intuitive and analytic thought provides us with a new approach to stochastic thinking. We had originally distinguished between two forms of stochastic thinking. We labeled thinking "stochastic", if subjects are "coping with stochastic situations and/or ... when the chance or probability concept is referred to, or stochastic models are applied.", p.4. While in the former, a subjective stochastic conceptualization is not necessarily needed, the latter does not have to involve stochastic situations (cf. Chapter 1). We will close this chapter by pointing out which new perspectives and also insights have been gained from the introduced framework.

Investigating the process and structure of stochastic thinking. We propose that the considerations in this chapter provide a differentiated analysis and understanding of the stages and characteristics of stochastic thinking that answers the demands presented in the "Guidelines for forthcoming research" given at the end of the Conclusions in Chapter 2. Let us exemplarily sketch what individuals are doing if they commence to cope with a situation or problem such as the Hit Parade problem. In this we will apply both the terminology introduced in this chapter and the results of Experiments A to D. According to Chapters 3 and 4, a crucial assumption which has to be fixed before such a description is provided is the mode of thought which the subject is believed to use in his or her processing. Let us assume an individual is preinitialized to the intuitive mode, then the following may hold:

linguistic one, for instance in his computational analysis of English uncertainty terms, FOX, 1985, distinguished possibility, plausibility, probability, doubt, and belief.

The framework allows for an integrated view of the different approaches to a modeling of inferential heuristics in stochastic thinking. Former psychological research approaches into conceptualization of cognitive strategies in probability judgment primarily dealt with a small number of judgmental heuristics (cf. KAHNEMAN, SLOVIC, & TVERSKY, 1982; BAR—HILLEL, 1983; WALLSTEN, 1983). From a theoretical perspective, the most severe criticism in this field of research has been that the different heuristics are defined side by side. We think that the proposed framework for the process and structure of information processing in stochastic thinking provides a basis for an integration of the classical heuristics, and furthermore, even offers a means of conceptualizing the differential status of the various heuristics.

For instance, the availability heuristic (cf. Chapter 2.3.; and also Comment a) on the Sensory System and the section on the Knowledge Base; both in Chapter 5.1.1.) is primarily based on processes which are localized in the initial phase of information processing, i.e., the encoding and retrieval, although availability may be considered as a very general concept, which may concern each activation process, for instance, any activation of heuristics, goals, or evaluative operations. The representativeness heuristic, as a global similarity matching heuristic [cf. Chapter 2.3.; Comment e) on the Heuristic Structure in Chapter 5.1.1.], has to be regarded as a typical, simple everyday heuristic. However, representativeness may also be determined by analytic levels. Within the framework, the causal schema has to be conceived as a more complex cognitive activity that not only involves the Heuristic Structure but presumably also often the Evaluative Structure, as the causality between entities is seldom present in the Knowledge Base, and is nearly always the result of an evaluative, comparative, and reconstructive procedure. These brief remarks reveal that, in decision research, essentially different processes that need to be conceptually separated have been labeled with one and the same term, namely heuristics.

Within the protocol analyses of Experiments B and C, further heuristics have been identified such as rough middling strategies between displayed numerical values or intuitive information weighting. The individualized normative solution from Chapter 2 may also be regarded as an element of the Heuristic Structure. Yet clearly it has to be perceived as an element of higher ordered heuristics, as we know from many psychological experiments (cf.

Chapter 1.) that BAYES' theorem is not an intuitive inferential guideline. Mathematical theorems, or statistical or probability theorems, tests, or inferences may also be embedded in the higher ordered heuristics.

Toward an investigation of the relation between the individual's concept field of probability and the heuristic structure in stochastic thinking. Mathematical theorems, statistical tests, algorithmic solution procedures, rules of logic, and knowledge about causal connections in medicine or other conceptual fields are two−sided. They possess a static, conceptual, knowledge−based side, and also an operative, active, dynamic heuristic side.

If we apply this more epistemological view in stochastic thinking, we have to hypothesize a connection between the stochastic conceptual elements that are activated and the heuristics that are applied.

In the preceding chapters, we have introduced a series of heuristics on a model level and by examples for both the intuitive and the analytic modes of thought. To some extent, we were able to show a connection to different interpretations of the probability concept. One example is information weighting, with subsequent rough information integration generated through estimation procedures, when referring to either the logical or the subjective probability concept (cf. Chapter 2.4.). A further example is the classical representativeness heuristic which is based on a global similarity judgment. In this example, a contingency with the subjective probability concept may be hypothesized. Clearly, we are still far from an elaborated theory about the interplay between the Heuristic Structure and the Knowledge Base, and more insight into the two single units is needed. However, some recent studies performed on strategically relatively easy and conceptually more unequivocal two−spinner−tasks (c.f. SCHOLZ & WASCHESCIO, 1986) allow us to anticipate a level of description of the individual's Heuristic Structure and Knowledge Base which will not only permit a formal precise description, but also a clarification of the interplay of the different entities of long−term storage.

Modes of thought, the framework, and progress in cognitive decision research. The preceding sections outlined the new perspectives and insights on stochastic thinking (for instance, an adequate understanding of the differential status of various heuristics) which have been achieved by referring to the framework of information processing. On the other hand, stochastic thinking and especially the research on probability judgment is a suitable subject to provide insight into the structure, the functional principles and the mutual dependency of the various cognitive units. But a third 'conceptual anchor' has been introduced and related to both the subject of stochastic thinking and the elaboration of the framework. The complementarity between the analytic and the intuitive modes of thought was introduced (cf. Chapter 2.) in order to categorize the qualitatively different strategies which were found in probability judgments on base−rate problems. The (interrater) reliability of the modes and the dissimilarity of the response distributions of these two modes have been demonstrated in two experiments. This chapter proposes a reformulation of the modes within the dynamics of the framework which provides a process view on intuitive and analytic cognitive activity in stochastic thinking.

Nevertheless, both decision making and stochastic thinking are "one of the pivots of cognition. Traditional psychological theories based on statistical concepts, and modern 'heuristic' theories, take an overly narrow view. Relationships that must exist between decision making and other cognitive capabilities like memory, reasoning, problem solving and language are ignored" (FOX, 1985). Not all these relationships have been dealt with in this text, nevertheless, we think that the approach via the modes of thought and the framework of the process and structure of human information processing, and its elaboration in the context of stochastic thinking, has accomplished a solid basis for a theoretical integration of problem solving and decision research. It is to be hoped that future work will be able to step down from the level of theoretical frameworks which have to be conceived as theory schemes, and formulate specific theories and models which will permit experimental plans for testing the validity of the theory, and also attain the necessary degree of precision that will permit the design and running of meaningful simulation studies which account for the organization and processing principles of human cognition.

REFERENCES

Abelson, R.P.: 1978. Invited address, Midwestern Psychological Association, Chicago. Quoted according to: Carver, C.S. & Scheier, M.F., Attention and Self — Regulation: A Control — Theory Approach to Human Behavior. New York: Springer

Abt, K.: 1983. Significance Testing of Many Variables. Neuropsychology, 9, 47 — 51

Aebli, H.: 1980. Denken: Das Ordnen des Tuns. Band I: Kognitive Aspekte der Handlungstheorie. Stuttgart: Klett — Cotta

Aebli, H.: 1981. Denken: Das Ordnen des Tuns. Bd. II: Denkprozesse. Stuttgart: Klett — Cotta

Ajzen, I.: 1977. Intuitive Theories of Events and the Effect of Base — Rate Information on Prediction. Journal of Personality and Social Psychology, 35(5), 303 — 314

Allwood, C.M.: 1982. Use of knowledge and error detection when solving statistical problems. In: Vermandel, A. (ed.), Proceedings of the sixth international conference for the psychology of mathematical education, Wilrijk (Belgium): Universitaire Instelling Antwerpen, 274 — 279

Allwood, C.M.: 1984. Error Detection Processes in Statistical Problem Solving. Cognitive Science, 8, 413 — 437

Allwood, C.M.: 1985. The Evaluation of Two Types of Knowledge at Different Points in Time after Study. In: Streefland, L. (ed.), Proceedings of the Ninth International Conference for the Psychology of Mathematics Education. Utrecht: State University of Utrecht, 453 — 458

Allwood, C.M. & Montgomery, H.: 1981. Knowledge and technique in statistical problem solving. European Journal of Science Education, 3(4), 431 — 450

Allwood, C.M. & Montgomery, H.: 1982. Detection of errors in statistical problem solving. Scandinavian Journal of Psychology, 23, 131 — 139

Anderson, J.R.: 1982. Acquisition of cognitive skill. Psychological Review, 89, 369 — 406

Anderson, J.R.: 1983. The architecture of cognition. Cambridge, Mass.: Harvard University Press

REFERENCES

APA (American Psychological Association): 1954. Quoted according to Weiner, B.: 1976. Theorien der Motivation, Stuttgart: Klett

Aschenbrenner, K.M.: 1977. Komplexes Wahlverhalten: Entscheidungen zwischen multiattributen Alternativen. In: Hartmann, K.D. & Koeppler, K. (eds.), Fortschritte der Marktpsychologie 1, 21−52

Asendorpf, J.: 1984. Lassen sich emotionale Qualitäten im Verhalten unterscheiden? Empirische Befunde und ein Dilemma. Psychologische Rundschau, 35(3), 125−135

Bar−Hillel, M.: 1975. The base−rate fallacy in subjective judgments of probability. Unpublished doctoral dissertation (in Hebrew). The Hebrew University of Jerusalem. Quoted according to Tversky, A. & Kahneman, D., 1979

Bar−Hillel, M.: 1980. The Base−Rate Fallacy in Probability Judgments. Acta Psychologica, 44, 211−233

Bar−Hillel, M.: 1983. The Base−Rate Fallacy Controversy. In: Scholz, R.W. (ed.), Decision Making under Uncertainty. Amsterdam: Elsevier (NorthHolland), 39−63

Bar−Hillel, M.: 1984. Representativeness and fallacies of probability judgment. Acta Psychologica, 55, 91−107

Bar−Hillel, M. & Falk, R.: 1982. Some teasers concerning conditional probabilities. Cognition, 11, 109−122

Bar−Hillel, M. & Fischhoff, B.: 1981. When do base rates affect predictions? Journal of Personality and Social Psychology, 41, 671−680

Bartlett, F.C.: 1932. Remembering. A Study in Experimental and Social Psychology. Cambridge: Cambridge University Press

Bastick, T.: 1982. Intuition, How We Think and Act. Chichester: Wiley

Baumann, K. & Sexl, U.: 1984. Die Deutungen der Quantentheorie. Braunschweig: Vieweg

Berkeley, D. & Humphreys, P.: 1982. Structuring Decision Problems and the Bias Heuristic. Acta Psychologica, 50, 201−252

Berne, E.: 1949. The nature of intuition. Psychiatric Quarterly, 23, 203−226

Bernstein, B.: 1977. Class, Codes and Control, Vol. 3. Towards a Theory of Educational Transmissions. London: Routledge & Kegan Paul

Beyth−Marom, R.: 1982. How probable is probable? Numerical translation of verbal probabilitity expressions. Journal of Forecasting, 1, 257−269

Beyth−Marom, R. & Arkes, H.R.: 1983. Being accurate but not necessarily Bayesian: comments on Christensen−Szalanski and Beach. Organizational Behavior and Human Performance, 31, 225−257

Beyth−Marom, R. & Fischhoff, B.: 1983. Diagnosticity and pseudodiagnosticity. Journal of Personality and Social Psychology, 45(6), 1185−1195

Beyth−Marom, R. & Lichtenstein, S.: in press. An Elementary Approach to Thinking under Uncertainty. Hillsdale, N.J.: Erlbaum

Biehler, R.: 1982. Explorative Datenanalyse. IDM−Materialien und Studien Vol. 23, Bielefeld: Universität Bielefeld

Birnbaum, M.H.: 1983. Base−Rates in Bayesian Inference: Signal Detection Analysis of the Cab Problem. American Journal of Psychology, 96(1), 85−94

Birnbaum, M.H. & Mellers, B.A.: 1983. Bayesian Inference: Combining Base Rates With Opinions of Sources Who Vary in Credibility. Journal of Personality and Social Psychology, 45(4), 792−804

Birnbaum, M.H. & Stegner, S. E.: 1981. Measuring the Importance of Cues in Judgment for Individuals: Subjective Theories of IQ as a Function of Heredity and Environment. Journal of Experimental Social Psychology, 17, 157−182

Bjork, R.A.: 1975. Short−term storage: the ordered output of a central processor. In: Restle, F., Shiffrin, R.M., Casstellan, N.S., Lindmann, H.R., & Pisoni, D.B. (eds.), Cognitive theory. Vol.1. Hillsdale, N.J.: Erlbaum, 151−171

Borgida, E. & Brekke, N.: 1981. The Base−Rate Fallacy in Attribution and Prediction. In: Harvey, J.H., Ickes, W.J., & Kidd, R.F., (eds.), New Directions in Attribution Research (Vol. 3). Hillsdale, N.J.: Erlbaum, 66−95

Borovcnik, M.: 1982. Case studies for an adequate understanding and interpretation of results by statistical inference. Paper presented at the "First International Conference on Teaching Statistics.", Sheffield, 8.−13.8. 1982

Borovcnik, M.: 1984. Der Problemkreis Bayessche Formel. In: mathematica didactica, 7, p. 207−224

Bortz, J.: 1985. Lehrbuch der Statistik, 2nd ed., Berlin: Springer

Box, G.E.P.: 1953. Non−normality and tests on variance. In: Biometrika, 40, 318−335

Box, G.E.P.: 1954. Some theorems on quadratic forms applied in the study of analysis of variance problems. II. Effect of inequality of variances in one−way classification. Annals of Mathematical Statistics, 25, 290−302

Brachmann, R.J.: 1979. On the epistemological status of semantic networks.

In: FinJler, N.V. (ed.), Associative networks: Representation and use of knowledge by computers. New York: Academic Press, 3−50

Braine, M.D.S.: 1978. On the Relation between the Natural Logic of Reasoning and Standard Logic. Psychological Review, 85(1), 1−21

Bratus, B.S.: 1978. The Mechanisms of Goal−Setting. Soviet Psychology, 3, 69−77

Brehmer, B.: 1980. In one word: not from experience. Acta Psychologica, 45, 223−241

Broadbent, D.E.: 1971. Decision and stress. New York: Academic Press

Brody, T.A.: 1983. Foundations for Quantum Mechanics. Results and Problems. Revist a Mexican de Fisica, 29(4), 461−507

Bromme, R., Bussmann, H., Heymann, H.W., Lorenz, J.H., Reiss, V., Scholz, R.W., & Seeger, F.: 1983. Methodological problems of object−adequate modelling and conceptualization of teaching, learning, and thinking processes related to mathematics. Bielefeld: IDM−University of Bielefeld, Occasional Paper 38, 1983. Printed in: 1983. Zweng, M. et al. (eds.), ICME 4. Boston: Birkhäuser, 468−469

Bromme, R. & Hömberg, E.: 1977. Psychologie und Heuristik. Darmstadt: Steinkopf

Bromme, R. & Steinbring, H.: 1981. Lokale und Globale Aspekte des Lehrerwissens. IDM Occasional Paper 8, Bielefeld: Universität Bielefeld

Bruner, J.S.: 1960. The Process of Education. Harvard: Harvard University Press

Bruner, J.S. & Clinchy, B.: 1966. Towards an disciplined intuition. In: Bruner, J. (ed.): Learning About Learning no. 15. Bureau of Research Monograph (Superintendent of Documents Catalog No. FS 5.212.: 12019)

Brunswik, E.: 1969. The conceptual framework of psychology. Chicago: University of Chicago Press

Bungard, W.: 1980. Die "gute" Versuchsperson denkt nicht. München: Urban und Schwarzenberg

Bunge, M.: 1962. Intuition and Science. Englewood Cliffs, N.J.: Prentice Hall

Caldwell, J. & Goldin, G.A.: 1979. Variables affecting word problem difficulty in school mathematics. In: Journal for Research in Mathematics Education, 10(5), 323−336

Carnap, R.: 1962. Logical Foundations of Probability. Chicago: University Press

Carroll, J.S. & Siegler, R.S.: 1977. Strategies for the use of base−rate information. Organizational Behavior and Human Performance, 19, 392–402

Carver, Ch.S. & Scheier, M.F.: 1981. Attention and Self–Regulation: A Control–Theory Approach to Human Behavior. New York: Springer

Chaiklin, S.: 1985. Beyond Inferencing: Some Cognitive Processes that Affect the Direction of Student Reasoning in Applying Physical–Science Beliefs. Paper presented at the Conference on Cognitive Processes in Student Learning. Lancester, U.K., July, 18–21

Chapman, R.H.: 1975. The Development of Children's Understanding of Proportion. Child Development, 46, 141–148

Chase, W.G. & Clark, H.H.: 1972. Mental operations in the comparison of sentences and pictures. In: Gregg, L.W. (ed.), Cognition in Learning and Memory. New York: Wiley

Chi, M.T.H., Glaser, R. & Rees, E.: 1982. Expertise in problem solving. In: R.J. Sternberg (ed.), Advances in the Psychology of Human Intelligence, Vol. 1., Hillsdale, N.J.: Erlbaum, 7–75

Christensen–Szalanski, J.J.J. & Beach, R.: 1982. Experience and the base–rate fallacy. Organizational Behavior and Human Performance, 29, 270–278

Christensen–Szalanski, J.J.J. & Beach, R.: 1983. Believing is not the same as testing: a reply to Beyth–Marom and Arkes. Organizational Behavior and Human Performance, 31, 258–261

Christensen–Szalanski, J.J.J. & Beach, R.: 1984. The Citation Bias: Fad and Fashion in the Judgment and Decision Literature. American Psychologist, 1, 75–78

Christensen–Szalanski, J.J.J. & Bushyhead, J.B.: 1981. Physicians use of probabilistic information in a real setting. Journal of Experimental Psychology: Human Perception and Performance, 7, 928–935.

Clark, M.S.: 1982. A Role for Arousal in the Link between Feeling States, Jugdments, and Behavior. In: Clark, M.S. & Fiske, S. (eds.), Affect and Cognition, Hillsdale, N.J.: Erlbaum, 263–288

Clark, M.S. & Isen, A.M.: 1982. Toward understanding the relationship between feeling states and social behavior. In: Hastorf, A. & Isen, A.M. (eds.), Cognitive Social Psychology. New York: Elsevier, 78–113

Clarkson, G.P.E.: 1962. Portfolio Selection: A Simulation of Trust Investment. Englewood Cliffs, N.J.: Prentice–Hall

Clinchy, B.: 1975. The role of intuition on learning. Today's Education, 64(2), 48–51

REFERENCES

Cohen, J.: 1964. Behavior in Uncertainty. New York: Basic Books
Cohen, J. & Hansel, C.E.M.: 1958. The nature of Decisions in Gambling. Acta Psychologica, 13, 357−370
Cohen, L.J.: 1977. The Probabable and the Provable. Oxford: Clarendon Press
Cohen, L.J.: 1979. On the Psychology of Prediction: Whose is the Fallacy? Cognition, 7, 385−407
Cohen, L.J.: 1981. Can Human Irrationality be Experimentally Demonstrated? The Behavioural and Brain Sciences, 4, 317−331
Collins, A.M. & Loftus, E.F.A.: 1975. Spreading−activation theory of semantic processing. Psychological Review, 82, 407−428
Collins, A.M. & Quillian, M.R.: 1969. Retrieval time from semantic memory. Journal of Verbal Learning and Verbal Behavior, 8, 240−247
Courtney, R.: 1968. Play, Drama and Thought. London: Cassell
Cox, J.R. & Griggs, R.A.: 1982. The effects of experience on performance in Wason's selection task. Memory and Cognition, 10, 496−502
Craik, F.I.M. & Lockhart, R.S.: 1972. Levels of processing: A framework for memory research. Journal of Verbal Learning and Verbal Behavior, 11, 671−684
Crott, H.W., Scholz, R.W., Ksiensik, M.I., & Popp, M.: 1983. Koalitionsentscheidungen und Aufteilungsverhalten in Drei−Personen−Spielen. Theoretische und experimentelle Untersuchungen zu Konflikt, Macht und Anspruchsniveau. Frankfurt: Lang
Crott, H.W. & Zuber, J.A.: 1983. Biases in Group Decision Making. In: Scholz, R.W. (ed.), Decision Making under Uncertainty. Amsterdam: North−Holland, 229−252
Dahlke, R. & Fakler, R.: 1982. Geometrical Probability − A Source of Interesting and Significant Applications of High School Mathematics. Mathematics Teacher, 736−745
Dawes, R.M.: 1975. The mind, the model, and the task. In: Restle, F., Shiffrin, R.M., Castellan, N.J., Lindman, H.R., & Pisoni, D.F. (eds.), Cognitive theory, Hillsdale, N.J.: Erlbaum, 119−129
Dawes, R.M.: 1979. The robust beauty of improper linear models in decision making. American Psychologist, 34, 571−582
De Finetti, B.: 1951. Recent Suggestions for the Reconciliations of Theories of Probability. In: Neymann, J. (ed.), Proceedings of the Second Berkeley Symposium on Mathematical Statistics and Probability. Berkeley: University of California, 217−226

REFERENCES

De Finetti, B.: 1972. Probability, Induction and Statistics. New York: Wiley

De Finetti, B.: 1974. The True Subjective Probability Problem. In: Stael von Holstein, C. – A. (ed.), The concept of probability in psychological experiments. Dordrecht – Holland: Reidel, 15 – 24

De Groot, A.D.: 1965. Thought and Choice in Chess. The Hague: Mouton

Dempster, A.P.: 1968. A generalization of Bayesian inference. Journal of the Royal Statistical Society, Series B, 30, 205 – 247

Diehl, J.M.: 1977. Varianzanalyse. Frankfurt: Fachbuchhandlung für Psychologie

Dinges, H.: 1977. Zum Umgang mit Aussagen in der Stochastik (Bemerkungen zum stochastischen Denken), unpublished manuscript, University of Frankfurt; quoted according to Steinbring, H.: 1980. Zur Entwicklung des Wahrscheinlichkeitsbegriffs – Das Anwendungsproblem in der Wahrscheinlichkeitstheorie aus Didaktischer Sicht. IDM – Materialien und Studien Vol. 18. Bielefeld: Universität Bielefeld

Dinges, H.: 1979. Report on Stochastics at Highschools in the Federal Republic of Germany. Unpublished manuscript: Frankfurt: University of Frankfurt

Dinges, H. & Rost, H.: 1982. Prinzipien der Stochastik. Stuttgart: Teubner

Dörner, D.: 1976/1979. Problemlösen als Informationsverarbeitung. Stuttgart: Kohlhammer, 1st and 2nd edition

Dörner, D.: 1981. Über Schwierigkeiten menschlichen Umgangs mit Komplexität. Psychologische Rundschau, 31(13), 163 – 179

Dörner, D.: 1984. Denken, Problemlösen und Intelligenz. Psychologische Rundschau, 35(1), 10 – 20

Dörner, D., Kreuzig, H.W., Reither, F., & Stäudel, T.: 1983. Vom Umgang mit Unbestimmtheit und Komplexität. Bern: Huber

Duncker, K.: 1935. Zur Psychologie des produktiven Denkens. Berlin: Springer

Earle, T.C.: 1971. Intuitive and Analytical Thinking in Learning and Interpersonal Learning. Unpublished paper. Eugene: Oregon Research Institute

Earle, T.C.: 1972. Intuitive and Analytical Thinking in Consistent and Inconsistent Multiple – Cue Learning Tasks. Unpublished PhD thesis, University of Oregon

Eddy, D.M.: 1982. Probabilistic reasoning in clinical medicine: Problems and Opportunities. In: Kahnemann, D., Slovic, P., & Tversky, A. (eds.), Judgment under uncertainty: Heuristics and biases. Cambridge: University Press, 249 – 267.

REFERENCES

Edwards, A.W.F.: 1972. Likelihood. Cambridge: University Press

Edwards, W.: 1983. Human Cognitive Capabilities, Representativeness, and Ground Rules for Research. In: P. Humphreys, O. Svenson, A. Vari (eds.), Analysing and Aiding Decision Processes. Amsterdam: Elsevier (North–Holland), 507–514

Edwards, W., Lindman, H., & Savage, L.J.: 1963. Bayesian statistical inference for psychological research. Psychological Review, 70, 193–242

Einhorn, H.J.: 1980. Learning from experience and suboptimal rules in decision making. In: Th.S. Wallsten (ed.), Cognitive Processes in Choice and Decision Behavior. Hillsdale, N.J.: Erlbaum, 1–17

Einhorn, H.J. & Hogarth, R.M.: 1981. Behavioral Decision Theory: Processes of Judgment and Choice. Annual Review of Psychology, 32, 53–88

Encyclopaedia Britannica: 1981. Micropaedia, 15–th edition, Vol IX, Chicago.

Ericsson, K.A. & Simon, H.A.: 1980. Verbal reports as data. Psychological Review, 87, 215–251

Ericsson, K.A. & Simon, H.A.: 1984. Protocol Analysis: Verbal Reports as Data. London: The MIT Press

Estes, W.K.: 1964. "Probability Learning". In: A.W. Melton (ed.), Categories of Human learning. New York: Academic Press, 89–128.

Estes, W.K.: 1976. The Cognitive Side of Probability Learning. Psychological Review, 83(1), 37–64

Fennema, E.: 1979. Woman and Girl in Mathematics – Equity in Mathematics Education. Educational Studies in Mathematics, 10, 389–401

Fiedler, K.: 1980. Kognitive Verarbeitung statistischer Information: Der "vergebliche Konsensus–Effekt". Zeitschrift für Sozialpsychologie, 11, 25–37

Fiedier, K.: 1985. Handlungssteuerung durch die Dialektik von Kognition und Emotion. Lecture held at the 2nd Tagung der Arbeitsgruppe Sozialpsychologie 15.–17.2.85, EWH–Landau

Fischbein, E.: 1975. The intuitive sources of probabilistic thinking in children. Dordrecht: Reidel

Fischbein, E.: 1982. Intuition and proof. For the Learning of Mathematics 3 (2), 9–25

Fischbein, E.: 1983. Intuition and Axiomatics in Mathematical Education. In: M. Zweng et al. (eds.), Proceedings of the 4th International Congress on Mathematical Education. Boston: Birkhäuser, 599–602

Fischhoff, B.: 1983a. Reconstructive Criticism. In: P. Humphreys, O. Svenson, & A. Vari (eds.): Analysing and Aiding Decision Processes. Amsterdam: Elsevier (North—Holland), 515—524

Fischhoff, B.: 1983b. Predicting Frames. Journal of Experimental Psychology: Learning, Memory & Cognition, 9, 103—116

Fischhoff, B. & Bar—Hillel, M.: 1984a. Diagnosticity and the Base—Rate Effect. Memory & Cognition, 12 (4), 402—410

Fischhoff, B. & Bar—Hillel, M.: 1984b. Focussing techniques: A shortcut to improving probability judgments? Organizational Behavior and Human Performance, 33, 175—194

Fischhoff, B. & Beyth—Marom, R.: 1983. Hypothesis Evaluation from a Bayesian Perspective. Psychological Review, 90(3), 239—260

Fischhoff, B., Slovic, P., & Lichtenstein, S.: 1978. Fault trees: Sensitivity of estimated failure probabilites to problem representation. In: Journal of Experimental Psychology: Human Perception and Performance, 4, 330—344

Fischhoff, B., Slovic, P., & Lichtenstein, S.: 1979. Subjective sensitivity analysis. Organizational Behavior and Human Performance, 23, 339—359

Fishbein, M.: 1967. The Prediction of Behaviors from Attitudinal Variables. In: Fishbein, M. (ed.): Readings in Attitude Theory and Measurement. New York: Wiley

Flavell, J.H.: 1963. The developmental psychology of Jean Piaget. Princeton, N.J.: Van Nostrand

Flavell, J.H.: 1979. Kognitive Entwicklung. Stuttgart: Klett Cotta. Engl.: Cognitive Development. Englewood Cliffs, N.J.: Prentice Hall

Fox, J.: 1985. Logic, language and decision making. Paper presented at the International Workshop on Modelling Cognition. Lancester, U.K.

Freudenthal, H.: 1961. Models in Applied Probability. In: Kazemeir, B.H. & Vuysje, D., (eds.): The Concept and the Role of the Model in Mathematics and Natural and Social Sciences. Dordrecht: Reidel, 78—88

Freudenthal, H.: 1973. Mathematik als pädagogische Aufgabe, Vol. 2, Stuttgart: Klett

Geissler, H.G.: 1980. Perceptual Representation of Information: Dynamic Frames of Reference in Judgment and Recognition. In: Klix, F. & Krause, B., (eds.), Psychological Research Humboldt—Universität 1960—1980, Berlin: Verlag der Wissenschaften, 53—85

Ginosar, Z. & Trope, Y: 1980. The effects of base rates and individuating information on judgments about another person. Journal of Experimental

Social Psychology, 16, 228—242

Glaser, R.; 1984. The Role of Thinking, American Psychologist, 39(2), 93 — 104

Goffman, E.: 1974. Frame Analysis. Cambridge, Mass.: Harvard University Press

Goldberg, L.R.: 1970. Man versus model of man: A rationale, plus some evidence, for a method of improving on clinical inferences. Psychological Bulletin, 73(6), 422—432

Goldin, G.A. & Mc Clintock, C.E. (eds.): 1979. Task Variables in Mathematical Problem Solving. Columbus: ERIC

Good, I.J.: 1959. Kinds of Probability. Science, 129(3347), 443—447

Gough, H.G.: 1962. Clinical vs. statistical prediction in psychology. In: Postmann, L. (ed.), Psychology in the making. New York: Knopf, 562—584

Green, D.R.: 1982a. Testing Randomness. Teaching Mathematics and its Applications, 1(3), 95—100

Green, D.R.: 1982b. A Survey of probability concepts in 3.000 pupils aged 11—16 years. In: Barnett, V. et al. (eds.). International Conference on Teaching Statistics. Sheffield: University Press, 766—783

Green, P.M. & Swets, J.A.: 1966. Signal Detection Theory and Psychophysics. New York: Wiley

Groner, R., Groner, U., & Bischof, W.F.: 1983. The Role of Heuristics in Models of Decision. In: Scholz, R.W. (ed.): Decision Making under Uncertainty. Amsterdam: Elsevier (North—Holland), 87—108

Hacking, I.: 1975. The emergence of probability. Cambridge: University Press

Hammond, K.R.: 1984. Unification of Theory and Research in Judgment and Decision Making. Manuscript. University of Colorado

Hammond, K., Hamm, R.M., Grassia, J., & Pearson, T.: 1983. Direct Comparison of Intuitive, Quasi—Rational, and Analytical Cognition. Research Report: Center for Research on Judgment and Policy Institute of Cognitive Science, University of Colorado at Boulder

Hammond, K.R., Hursch, C.J., & Todd, F.J.: 1964. Analyzing the components of clinical inference. Psychological Review, 71, 438—456

Hammond, K.R., McClelland, G.H., & Mumpower, J.: 1980. Human Judgment and Decision Making: Theories, Methods, and Procedures. New York: Hemisphere/Praeger

Harten, G. v., Jahnke, H.N., Mormann, Th., Otte, M., Seeger, F., Steinbring, H., & Stellmacher, H. (eds.): 1985. Funktionsbegriff und

REFERENCES

funktionales Denken. Köln: Aulis

Harten, G. v. & Steinbring, H.: 1984. Stochastik in der Sekundarstufe I. Köln: Aulis

Hausmann, Chr.: 1985. Iterative and Recursive Modes of Thinking in Mathematical Problem Solving Processes. In: Streefland, L. (ed.) Proceedings of the Ninth International Conference for the Psychology of Mathematics Education, Utrecht: State University of Utrecht, 18—24

Hawkins, A.S. & Kapadia, R.: 1984. Children's Conceptions of Probability — A Psychological and Pedagogical Review. Educational Studies in Mathematics, 15, 349—477

Hayes, J.R.: 1982. Issues in protocol analysis. In: G.U. Ungson & Braunstein, D.N. (eds.), Decision Making: An Interdisciplinary Inquiry. Boston: Kent, 61—77

Heckhausen, H.: 1980. Motivation und Handeln. Berlin: Springer

Heitele, D.: 1976. Didaktische Ansätze zum Stochastikunterricht. In: Grundschule und Förderstufe, unpublished PhD Dissertation, Dortmund

Herrmann, Theo: 1982. Über begriffliche Schwächen kognitivistischer Kognitionstheorien: Begriffsinflation und Akteur—System—Kontamination. Sprache & Kognition, 1, 3—14

Hershey, J.C. & Schoemaker, P.J.H.: 1980. Risk taking and problem context in the domain of losses: An expected utility analysis. Journal of Risk and Insurance, 47(33), 111—132

Hoemann, H.W. & Ross, B.M.: 1982. Children's Concept of Chance and Probability. In: Brainerd, Ch.J. (ed.): Children's Logical and Mathematical Cognition. New York: Springer, 93—122

Hörz, H., Löther, R., & Wollgast, S. (eds.): 1978. Wörterbuch Philosophie und Naturwissenschaften, Berlin: Dietz

Hoffmann, J. & Klimesch, W.: 1984. Die semantische Kodierung von Wörtern und Bildern. Sprache & Kognition, 3(1), 1—25

Hoffmann, J. & Klix, F.: 1977. Zur Prozeßcharakteristik der Bedeutungserkennung bei sprachlichen Reizen. Zeitschrift für Psychologie, 185, 315—368

Hogarth, R.M.: 1980. Judgment and Choice: The Psychology of Decision. Chicester: Wiley

Hogarth, R.M. & Makridakis, S.: 1981. Forecasting and Planing: An Evaluation. Management Science, 27(2), 115—138

Holt, D.L.: 1984. Auditors and Base—Rates revisited. Unpublished manuscript: School of Management, University of Minnesota

Hoppe, F.: 1931. Erfolg und Mißerfolg, Psychologische Forschung, 14, 1−62

Howell, W.A.: 1972. Compounding Uncertainty from Internal Sources. Journal of Experimental Psychology, 95(1), 6−13

Huber, B. & Huber, O.: 1984. The Development of the Concept of Qualitative (Comparative) Subjective Probability. Forschungsbericht Nr. 43, Freiburg/Schweiz: Universität Freiburg

Huber, O.: 1982. Entscheiden als Problemlösen. Bern: Huber

Huber, O.: 1983. The information presented and actually processed in a decision task. In: Humphreys, P., Svenson, O., & Vari, A. (eds.): Analysing and Aiding Decision Processes, Amsterdam, North−Holland, 441−454

Huff, D.: 1959. How to take a chance. Harmondsworth: Penguin Books

Humphreys, P. & Berkeley, D.: 1983. Problem Structuring Calculi and Levels of Knowledge Representation in Decision Making. In: Scholz, R.W. (ed.): Decision Making under Uncertainty. Amsterdam: Elsevier (North−Holland), 121−158

Husen, T. (ed.): 1967. International Study of Achievement in Mathematics. Hawthorn: Acer

Hussy, W.: 1983. Komplexe menschliche Informationsverarbeitung: das SPIV−Modell. Sprache & Kognition, 2, 47−62

Hussy, W.: 1984. Denkpsychologie. Band 1. Stuttgart: Kohlhammer

Irle, M.: 1975. Lehrbuch der Sozialpsychologie. Göttingen: Hogrefe

Izard, C.E.: 1977/1981. Human emotions. New York: Plenum Press. German: Die Emotion des Menschen. Weinheim: Beltz

Johnson, J.R. & Finke, R.A.: 1985. The base−rate fallacy in the context of sequential categories. Memory & Cognition. 13(1), 63−73

Johnson−Laird, P.N.: 1984. A Computational Analysis of Conciousness. Cognition and Brain Theory, 6(4), 499−508

Johnson−Laird, P.N. & Wason, P.C.: 1977. A theoretical analysis of insight into a reasoning task. In: P.N. Johnson−Laird & P.C. Wason (eds.), Thinking. Cambridge: Cambridge University Press, 143−157

Jungermann, J.: 1983. The Two Camps on Rationality. In: Scholz, R.W. (ed.), Decision Making under Uncertainty. Amsterdam: North−Holland, 63−86

Kahneman, D., Slovic, P., & Tversky, A.: 1982. Judgment under Uncertainty. Heuristics and Biases. Cambridge: Cambridge University Press

Kahneman, D. & Tversky, A.: 1972. Subjective Probability: A Judgment of Representativeness. Cognitive Psychology, 3, 430−454

Kahneman, D. & Tversky, A.: 1973. On the Psychology of Prediction. Psychological Review, 80, 4, 237–251

Kahneman, D. & Tversky, A.: 1979a. Intuitive Prediction: Biases and Corrective Procedures. TIMS Studies in the Management Science, 12, 313–327

Kahneman, D. & Tversky, A.: 1979b. On the interpretation of intuitive probability: A reply to Jonathan Cohen. Cognition, 7, 409–411

Kahneman, D. & Tversky, A.: 1979c. Prospect theory: An analysis of decision under risk. Econometrica, 47, 263–291

Kahneman, D. & Tversky, A.: 1982a. Variants of uncertainty. In: Kahneman, D., Slovic, P., & Tversky, A. (eds.), Judgment under Uncertainty: Heuristics and Biases. Cambridge: Cambridge University Press, 493–508

Kahneman, D. & Tversky, A.: 1982b. On the study of statistical intuitions. In: Kahneman, D., Slovic, P., & Tversky, A. (eds.): Judgment under Uncertainty: Heuristics and Biases. Cambridge: Cambridge University Press, 493–508

Karmiloff–Smith, A.: 1982. Modifications in children's representational systems and levels of accessing knowledge. In: Gelder, B. de (ed.): Knowledge and Representation. London: Routledge & Kegan Paul, 65–79

Kassin, S.M.: 1979. Consensus Information, Prediction, and Causal Attribution: A Review of the Literature and Issues. Journal of Personality and Social Psychology, 37, 1966–1981

Kemnitz, W.: 1976. Grundfragen der subjektivistischen Wahrscheinlichkeitskonzeption. Unpublished PhD Dissertation. Universität Konstanz

Keynes, J.M.: 1973/1921. A Treatise on Probability. London: MacMillan Press

Kintsch, W.: 1982. Memory for Text. In: Flammer, A. & Kintsch, W. (ed.), Discourse Processing, Amsterdam: North–Holland, 186–204

Klieme, E.: 1986. Bildliches Denken als Mediator für Geschlechtsunterschiede beim Lösen mathematischer Probleme. In: Steiner, H.–G. (ed.), Grundfragen der Entwicklung mathematischer Fähigkeiten. Köln: Aulis, 133–151

Klix, F.: 1971. Information und Verhalten. Huber: Bonn 1971

Klix, F.: 1980. General Psychology and the Investigation of Cognitive Processes. In: Klix, F. und Krause, B. (eds.): Psychological Research. Humboldt–Universität, Berlin: Verlag der Wissenschaften, 11–27

Kraatz, W.: 1978. Stochastik. In: Wörterbuch Philosophie und Naturwissenschaften, Hörz, H., Löther, R., & Wollgast, S. (eds). Berlin: Dietz, 861−865

Krantz, D.H., Luce, R.D., Suppes, P., & Tversky, A.: 1971. Foundations of measurement. Vol. 1. New York: Academic Press

Kruglanski, A.W.: 1982. Kognitive Sozialpsychologie: Eine Betrachtung zum kognitiven Pluralismus und Irrationalismus. Zeitschrift für Sozialpsychologie, 13, 150−162

Kuhl, J.: 1983a. Emotion, Kognition und Motivation: I. Auf dem Wege zu einer systemtheoretischen Betrachtung der Emotionsgenese. Sprache & Kognition, 4, 1−27

Kuhl, J.: 1983b. Emotion, Kognition und Motivation: II. Die funktionale Bedeutung der Emotionen für das problemlösende Denken und für das konkrete Handeln. Sprache & Kognition, 4, 228−253.

Kyburg, H.E.: 1970. Probability and Inductive Logic. London: Collier−Macmillan

Kyburg, H.E.: 1983. Rational Belief. The Behavioral and Brain Sciences, 6, 231−273

La Breque, M.: 1980. On making sounder judgments. Psychology Today, 6, 33−42

Laplace, P.S.: 1932. Essai Philosophique sur les Probabilite. Paris: Bachlier (5. Auflage). German: Philosophischer Versuch über die Wahrscheinlichkeit. Leipzig: Akademische Verlagsanstalt

Larkin, J., Mc Dermott, J., Simon, D.P., &Simon, H.A.: 1980. Expert and novice performance in solving physics problems. Science, 208, 1335−1342

Laugwitz, D.: 1979. Mathematikunterricht und Lehrerausbildung aus der Sicht einer technischen Hochschule. Lecture held at the DMV−Congress, Hamburg, 9.1.1979

Lee, W.: 1977. Psychologische Entscheidungstheorie. Weinheim: Beltz Engl.: 1971. Decision theory and human behavior. New York: Wiley

Lehmann, E.L.: 1975. Non−parametrics: Statistical methods based on ranks. San Franzisco: Holden−Day

Lewin, A.Y.: 1982. The state of the art in decision making: In integration of the issues. In: G.U. Ungson & D.N. Braunstein (eds.), Decision Making: An Interdisciplinary Inquiry. Boston: Kent, 312−316

Lewin, K.: 1926. Vorsatz, Wille und Bedürfnis. Psychologische Forschung, 7, 330−385

Lewin, K.: 1946. Verhalten und Entwicklung als eine Funktion der Gesamtsituation. Feldtheorie in den Sozialwissenschaften. Bonn: Huber, 271–329

Lichtenstein, S.: 1982. Commentary on Hayes' Paper. In: Ungson, G.U. & Braunstein, D.N. (eds.), Decision Making: An Interdisciplinary Inquiry. Boston: Kent, 83–85

Lichtenstein, S.: 1984. Personal Communication

Lichtenstein, S., Fischhoff, B., & Phillips, L.D.: 1980. Calibration of probabilities: The state of the art to 1980. In: Kahneman, D., Slovic, P. & Tversky, A., Judgment under Uncertainty: Heuristics and Biases. Cambridge: Cambridge University Press, 306–334

Lichtenstein, S. & Mc Gregor, D.: 1984. Base–Rate Instruction. Unpublished Manuscript, Eugene: Decision Research

Lichtenstein, S., Slovic, P., Fischhoff, B., Layman, M., & Combs, B.: 1978. Judged frequence of lethal events. Journal of Experimental Psychology: Human Learning and Memory, 4, 551–578

Lienert, G.A.: 1969. Testaufbau und Testanalyse. Weinheim: Beltz

Lindsay, P.H. & Norman, P.A.: 1977. Human Information Processing. 2nd ed., New York: Academic Press

Lisch, R. & Kriz, J.: 1978. Grundlagen und Modelle der Inhaltsanalyse. Hamburg: Rowohlt

Lompscher, H.J. (ed.): 1972. Theoretische und experimentelle Untersuchungen zur Entwicklung geistiger Fähigkeiten. Berlin: Volk und Wissen

Lopes, L.L.: 1982. Doing the impossible: A note on induction and the experience of randomness. Journal of Experimental Psychology, 8, 626–637

Lopes, L.L.: 1984. Performing competently. The Behavioral and Brain Sciences, 4, 343–344

Lusted, L.B.: 1968. Introduction to medical decision making. Springfield: Thomas

Lusted, L.B.: 1976. Clinical decision making. In: de Dombal, F.T. & Gremy, F. (eds.), Decision making and medical care. Dordrecht: Reidel, 77–98

Lyon, D. & Slovic, P.: 1976. Dominance of Accuracy Information and Neglect of Base–Rates in Probability Estimation. Acta Psychologica, 40, 287–298

Mai, N. & Hachmann, E.: 1977. Anwendung des Bayes Theorems in der medizinischen Diagnostik – Eine Literaturübersicht. METAMED, 1, 161–205

Mandler, G.: 1975. Mind and emotions. New York: Wiley

REFERENCES

Mandl, H., Spada, H., Albert, D., Dörner, D., & Möbus, C.: 1984. Antrag auf die Einrichtung eines Schwerpunktprogrammes "Wissenspsychologie" bei der DFG. Tübingen/Freiburg

Manis, M., Dovalina, J., Avis, N.E., & Cardoze, S.: 1980. Base rates can affect individual predictions. Journal of Personality and Social Psychology, 38, 231−248

May, R.: 1986a. Overconfidence as a Result of Incomplete and Wrong Knowledge. In: Scholz, R.W. (ed.), Current Issues in West German Decision Research. Frankfurt: Lang

May, R.: 1986b. Realismus von Wahrscheinlichkeiten. Frankfurt: Lang

Meehl, P.: 1954. Clinical vs. statistical prediction: A theoretical analysis and a review of the evidence. Minneapolis: University of Minnesota Press

Meehl, P. & Rosen, A.: 1955. Antecedent probability and the efficiency of psychometric signs, patterns, or cutting scores. Psychological Bulletin, 52, 194−216

Meyer, W.−U.: 1983. Emotionstheorien, Attributionsthereotische Ansätze. In: Euler, H.A. & Mandl, H. (eds.), Emotionspsychologie. München: Urban & Schwarzenberg, 80−85

Minsky, M.A.: 1975. A framework for representing knowledge. In: Winston, P. (ed.), The Psychology of Computer Vision. New York: McGraw Hill, 211−277

Mises, A. v.: 1928. Wahrscheinlichkeit, Statistik und Wahrheit. Wien: Springer

Montgomery, H. & Svenson, O.: 1983. A Think Aloud Study of Dominance Structuring in Decision Processes. In: Tietz, R. (ed.), Aspiration Levels in Bargaining and Economic Decision Making. Berlin: Springer, 366−383

Newell, A. & Simon, H.A.: 1972. Human problem solving. Englewood Cliffs, N.J.: Prentice−Hall

Neisser, U.: 1963. The multiplicity of thought. British Journal of Psychology, 54, 1−14

Neisser, U.: 1976. Cognition and reality. San Francisco: Freeman

Nisbett, R.E. & Borgida, E.: 1975. Attribution and the psychology of prediction. Journal of Personality and Social Psychology, 35, 613−634

Nisbett, R.E., Borgida, E., Crandell, R., & Reed, H.: 1976. Popular induction: Information is not necessarily informative. In: Caroll, J.S. & Payne, J.W. (eds.), Cognition and social behavior. Hillsdale, N.J.: Erlbaum, 227−236

Nisbett, R.E., Krantz, D.H., Jepson, C., & Kunda, Z.: 1983. The Use of Statistical Heuristics in Everyday Inductive Reasoning. Psychological Review, 90(4), 339–363

Nisbett, R.E. & Ross, L.: 1980. Human Inference: Strategies and Shortcomings of Social Judgment. Englewood Cliffs, N.J.: Prentice–Hall

Nisbett, R.E. & Wilson, T.D.: 1977. Telling more than we can know. Verbal reports on mental processes. Psychological Review, 84, 231–259

Norman, D.A. & Rumelhart, D.E.: 1975. Explorations in Cognition. San Francisco: Freeman

Oesterreich, R.: 1981. Handlungsregulation und Kontrolle. München: Urban & Schwarzenberg

Olson, C.L.: 1976. Some apparent violations of the representativeness heuristic in human judgment. Journal of Experimental Psychology: Human Perception and Performance, 2, 599–608

Ornstein, R.: 1976. Die Psychologie des Bewußtseins. Frankfurt/M.

Oswald, M. & Gadenne, V.: 1984. Wissen, Können und künstliche Intelligenz (Eine Analyse der Konzeption des deklarativen und prozeduralen Wissens). Sprache & Kognition, 3(3), 173–184

Otte, M. & Jahnke, H.N.: 1982. Complementary of theoretical terms – ratio and proportion as an example. Conference on function, Report 1, SLO, Enschede, 2, 97–113

Owen, G.: 1972. Spieltheorie. Heidelberg: Springer

Paivio, A.: 1971. Imagery and Verbal Processes. New York: Holt, Rinehart and Winston

Paivio, A.: 1977. Images, propositions and knowledge. In: Nicolas, J.M. (ed.), Images, perception and knowledge. Dordrecht: Reidel, 47–71

Payne, J.W., Braunstein, M.L., & Carroll J.S.: 1978. Exploring Predecisional Behavior: An Alternative Approach to Decision Research. Organizational Behavior and Human Performance, 22, 17–44

Perner, J.: 1979. Young Children's Bets in Probabilistic Tasks Involving Disjoint and Part–Whole related Events. Archives de Psychologie, 47, 131–149

Peters, J.T., Hammond, K.R., & Summers, D.A.: 1974. A Note on Intuitive vs. Analytic Thinking. Organizational Behavior and Human Performance 12, 125–131

Peterson, C.R. & Beach, L.R.: 1967. Man as an intuitive statistician. Psychological Bulletin, 68, 29–46.

Pfeiffer, H.: 1981. Zur sozialen Organisation von Wissen im Mathematikunterricht — eine erziehungssoziologische Analyse von Lehreraussagen. IDM—Materialien und Studien Vol. 21, Bielefeld: Universität Bielefeld

Phillips, L.D.: 1983. A theoretical perspective on heuristics and biases in probabilistic thinking. In: Humphreys, P., Svenson, O., & Vari, A., Analysing and aiding Decision Processes. Amsterdam: Elsevier (North Holland), 525—543

Piaget, J.: 1972. Die Entwicklung des Erkennens. I: Das mathematische Denken. Stuttgart: Klett

Piaget, J. & Inhelder, B.: 1951/1975. La genese de l'idee de hasard chez l'enfant. Paris: Presses Universitaires de France. Translated as: The Origin of the Idea of Chance in Children. New York: Norton

Pitz, G.: 1980. The Very Guide of Life: The Use of Probabilistic Information for Making Decisions. In: Wallsten, T.S. (ed.), Cognitive Processes in Choice and Decision Behavior. Hilldale, N.J.: Erlbaum, 77—87

Poincare, H.: 1929/1969. Intuition and Logic in Mathematics. The Mathematics Teacher, 1969, 205—212; reprinted from Poincare, H.: 1929. The Foundations of Science, New—York: The Science Press, 210—222

Pollatsek, A., Konold, C.E., Well, A.D., & Lima, S.D.: 1984. Beliefs Underlying Random Sampling. Memory and Cognition, 12, 395 — 401

Polya, S.: 1949. Schule des Denkens. Vom Lösen mathematischer Probleme. Bern: Francke; Engl.: 1945. How to solve it. Princeton: University Press

Posner, M.I.: 1982. Protocol analysis and human cognition. In: Ungson, G.U. & Braunstein, D, (eds.), Decision Making: An Interdisciplinary Approach. Boston: Kent, 78—82

Puri, M.L. & Sen, P.K.: 1971. Non—parametric methods in multivariate— analysis. New York: Wiley

Rapoport, A. & Wallsten, T.S.: 1972. Individual Decision Behavior. Annual Review of Psychology, 131—176

Reichenbach, H.: 1925. Wahrscheinlichkeitslehre. Leiden: Sijthoff

Reiss, V.: 1979. Zur theoretischen Einordnung von Sozialisationsphänomenen im Mathematikunterricht. Zeitschrift für Pädagogik, 2, 275—289

Renyi, A.: 1969. Briefe über die Wahrscheinlichkeit. Basel: Birkhäuser

Riemer, W.: 1981. Heuristische Begründungen und Modelle zu neutralen Begriffen und Sätzen der Stochastik. Wuppertal: Dissertation

Ritsert, J.: 1972. Inhaltsanalyse und Ideologiekritik. Ein Versuch über kritische Sozialforschung. Frankfurt: Athenäum Fischer

Roach, D.A.: 1979. The effect of conceptual style preference, related cognitive variables and sex on achievement in mathematics. British

Journal of Educational Psychology, 49, 79−82

Rosch, E.H. & Mervis, C.B.: 1975. Family resemblance: studies in the internal structure of categories. Cognitive Psychology, 7, 573−605

Rosier, M.: 1980. Changes in Secondary School Mathematics in Australia 1964−1978. Victoria: Acer

Ross, B.M.: 1966. Probability Concepts in Deaf and Hearing Children. Child Development, 37, 917−927

Ross, B.M.: & De Groot, J.F.: 1982. How Adolescents Combine Probabilities. The Journal of Psychology, 110, 75−90

Rothe, J., Seifert, R., & Timpe, K.−P.: 1980. Psychological Investigations on Receiving and Processing of Information during Work in Automated Industry. In: Klix, F. & Krause, B. (eds.), Psychological Research, Humboldt−Universität 1960−1980. Berlin: Verlag der Wissenschaften, 202−224

Russel, J.A.: 1980. A Circumplex Model of Affect. Journal of Personality and Social Psychology, 39(6), 1161−1178

Sahu, A.R.: 1983. On some Educational and Psychological Aspects of Problem Solving. International Journal of Mathematical Education in Science and Technology, 14(5), 555−563

Sauermann, H. & Selten, R.: 1962. Anspruchsniveauanpassungstheorie der Unternehmung. Zeitschrift für die gesamte Staatswissenschaft, 118, 577−597

Savage, L.J.: 1971. Elicitation of Personal Probalities and Expectations. Journal of the American Statistical Association, 66(20), 783−801

Sawyer, J.: 1966. Measurement and prediction, clinical and statistical. Psychological Bulletin, 66, 178−200

Schachter, S.: 1964. The interaction of cognitive and physiological determinants of emotional state. In: Berkowitz, L. (ed.), Advances in Experimental Social Psychology (Vol. I). New York: Academic Press, 49−80

Schachter, S. & Singer, J.: 1962. Cognitive, social, and physiological determinants of emotional state. Psychological Review, 69, 379−399

Schaefer, R.E.: 1985. Denken. Informationsverarbeitung, mathematische Modelle und Computersimulation. Berlin: Springer

Schank, R.C.: 1982. Depth of knowledge. In: Gelder, B. de (ed.): Knowledge and Representation. London: Routledge & Kegan Paul

Schildkamp−Kündiger, E.: 1980. Mathematics and gender. In: H.−G. Steiner (ed.), Comparative Studies of Mathematics Curricula − Change and Stability 1960−1980. IDM−Materialien und Studien Vol. 19,

Bielefeld: Universität Bielefeld, 601 – 622

Schildkamp – Kündiger, E.: 1982. An international review of gender and mathematics. Ohio: ERIC, Ohio State University

Schneider, W. & Scheibler, D.: 1983. Probleme und Möglichkeiten bei der Bewertung von Clusteranalyse – Verfahren. I. Ein Überblick über einschlägige Evaluationsstudien. Psychologische Beiträge, 25, 208 – 237

Schoemaker, P.J.H. & Kunreuther, H.C.: 1979. An experimental study of insurance decisions. Journal of Risk and Insurance, 46, 603 – 618

Schoenfeld, A.H.: 1983. Episodes and Executive Decisions in Mathematical Problem – Solving. In: Lesh, R. & Landau, M. (eds.), Acquisition of Mathematics, Concepts and Processes, London: Academic Press, 345 – 395

Scholz, R.W.: 1980a. Dyadische Verhandlungen. Eine theoretische und experimentelle Untersuchung von Vorhersagemodellen. Meisenheim am Glan: Hain

Scholz, R.W.: 1980b. Berufliche Fallstudien zum Stochastischen Denken. In: Beiträge zum Mathematikunterricht, Hannover: Schroedel 298 – 302

Scholz, R.W.: 1981. Stochastische Problemaufgaben: Analysen aus didaktischer und psychologischer Perspektive. IDM – Materialien und Studien Vol. 23, Bielefeld: Universität Bielefeld

Scholz, R.W.: 1983a. Introduction to Decision Making under Uncertainty: Biases, Fallacies, and the Development of Decision Making. In: Scholz, R.W. (ed.), Decision Making under Uncertainty. Amsterdam: North-Holland, 3 – 18

Scholz, R.W. (ed.): 1983b. Decision Making under Uncertainty. Amsterdam: Elsevier (North – Holland)

Scholz, R.W.: 1983c. Methodological problems of object – adequate learning, and thinking processes related to mathematics. In: Zweng, M. et al. (eds.), ICME 4, Boston: Birkhäuser, 468 – 469

Scholz, R.W.: 1983d. Changes in secondary school mathematics in Australia. Rosier, M.J., Rezension. ZDM 15(1), 44 – 47

Scholz, R.W. & Bentrup, A.: 1984. Age and Weight in Using Base – Rates. IDM – Occasional Paper 53, Bielefeld: Universität Bielefeld

Scholz, R.W. (ed.): 1986. Current Issues in West German Decision Research. Frankfurt: Lang

Scholz, R.W. & Köntopp, M.B.: Der Einfluß von Aufgabenrahmung, Art der Protokollierung und schulischer Sozialisation auf den Bearbeitungsmodus bei konjunktiven und disjunktiven Wahrscheinlichkeitsaufgaben, in prep.

Scholz, R.W., Seydel, U., Rechbauer, W., & Wentz, S.: 1983. Der Zusammenhang zwischen Verkehrsauffälligkeit, Persönlichkeitsfaktoren und Verhalten in Konfliktspielen. Zeitschrift für experimentelle und angewandte Psychologie, 30(2), 273–298

Scholz, R.W. & Waller, M.: 1983. Conceptual and Theoretical Issues in Developmental Research on the Acquisition of the Probability Concept. In: Scholz, R.W. (ed.), Decision Making under Uncertainty. Amsterdam: North–Holland, 291–312

Scholz, R.W. & Waschescio, R.: 1986. Childrens' Cognitive Strategies in Two–Spinner Roulette Tasks. In: Proceedings of the Tenth International Conference for the Psychology of Mathematics Education. London, 463–468

Schrage, G.: 1984. Stochastische Trugschlüsse. mathematica didactica, 7(3), 3–35

Schwartz, G.E. & Weinberger, D.A.: 1980. Pattern of Emotional Responses to Affective Situations: Relations Among Happiness, Sadness, Anger, Fear, Depression and Anxiety. Motivation and Emotion, 4(2), 151–191

Schwartz, G.E., Weinberger, D.A., & Singer, J.A.: 1981. Cordiovascular Differenciation of Happiness, Sadness, Anger, and Fear Following Imagery and Exercise. Psychosomatic Medicine, 43(4), 343–364

Seiler, T.B.: 1984. Was ist eine "konzeptuell akzeptable Kognitionstheorie"? Anmerkungen zu den Ausführungen von Theo Herrmann: Über begriffliche Schwächen kognitivistischer Kognitionstheorien. Sprache & Kognition, 2, 87–101

Selten, R.: 1983. Towards a theory of limited rationality. Some remarks on the symposium's impact on social economic theory. In: Scholz, R.W. (ed.), Decision Making under Uncertainty. Amsterdam: North Holland, 409–413

Selz, O.: 1913. Über die Gesetze des geordneten Denkverlaufs. Stuttgart: Speemann

Selz, O.: 1922. Zur Psychologie des produktiven Denkens und des Irrtums. Bonn: Cohen

Shafer, G.A.: 1976. A mathematical theory of evidence. Princeton, N.J.: Princeton University Press

Shafer, G.A. & Tversky, A.: 1985. Languages and Designs for Probability Judgment. Cognitive Science 9, 309–339

Shanteau, J.: 1980. The Concept of Weight in Judgment and Decision Making: A Review and Some Unifying Proposals. Research Report No.

228. Institute of Behavioral Science. University of Colorado

Shubik, M.: 1981. Personal communication during a lecture held at Bielefeld University

Shulman, L.S. & Elstein, A.S.: 1975. Studies of Problem Solving, Judgment, and Decision Making: Implications for Educational Research. In: Kerlinger, F.N. (ed.), Review of Research in Education, 3, 3–42

Siegel, S.: 1976. Nicht–parametrische Methoden. Frankfurt: Psychologische Verlagsbuchhandlung. Engl.: 1956. Non–parametric statistics for the behavioral sciences. New York: Mc Graw Hill

Simon, H.A.: 1955. A behavioral model of rational choice. Quarterly Journal of Economics, 69, 99–118

Simon, H.A.: 1958. Models of man. New York: Wiley

Simon, H.A.: 1979. Information processing models of cognition. Annual Review of Psychology, 30, 363–396

Sinz, R.: 1978. Gehirn und Gedächtnis. Stuttgart: Fischer

Skemp, R.R.: 1971. The Psychology of Learning Mathematics. Harmondsworth: Penguin

Slovic, P.: 1983. Framing Effects: Implications for Risk Preference, Risk Assesment, and Risk Communication. Paper presented at the Nato–ASI, Les Arcs, France

Slovic, P., Fischhoff, B., & Lichtenstein, S.: 1977. Behavioral decision theory. Annual Review of Psychology, 28, 1–39

Slovic, P., Fischhoff, B., & Lichtenstein, S.: 1982. Response mode, framing, and information–processing effects in risk assessment. In: Hogarth, R. (ed.), New Directions for Methodology of Social and Behavioral Science: Question Framing and Response Consistency, No. 11. San Francisco: Jossey–Bass, 21–36

Slovic, P. & Lichtenstein, S.: 1971. Comparison of Bayesian and Regression Approaches to the Study of Information Processing in Judgment. Organizational Behavior and Human Performance, 6, 649–744

Smith, E.E., Shoben, E.J., & Rips, L.J.: 1974. Structure and process in semantic memory. A featural model for semantic decisions. Psychological Review, 81, 214–241

Späth, H.: 1977. Cluster – Analyse – Algorithmen zur Objektklassifizierung und Datenreduktion. München: Oldenbourg

Steinbring, H.: 1980. Zur Entwicklung des Wahrscheinlichkeitsbegriffs: Das Anwendungsproblem in der Wahrscheinlichkeitstheorie aus didaktischer Sicht. IDM–Materialien und Studien Bd. 18, Bielefeld: Universität

Bielefeld

Steinbring, H. & Harten, G. v.: 1982. Learning from experience — Bayes Theorem: A Model for Stochastic Learning Situation? In: Barnett, V. et al., ICOTS—Proceedings, Sheffield, 701—713

Steiner, G.: 1980. Visuelle Vorstellungen beim Lösen von elementaren Problemen. Zur Frage nach dem Wesen der Vorstellungen. Stuttgart: Klett—Cotta

Steiner, H.—G.: 1984. Mathematisch—naturwissenschaftliche Bildung — kritisch—konstruktive Fragen und Bemerkungen zum Aufruf einiger Fachverbände. In: Reiss, M. & Steiner, H.G., (eds.), Mathematikkenntnisse, Leistungsmessung, Studierfähigkeit. Köln: Aulis, 5—59

Steinhausen, D. & Langer, K.: 1977. Clusteranalysen: Einführung in Methoden und Verfahren der automatischen Klassifikation, Berlin: De Gruyter

Stelzl, I.: 1982. Fehler und Fallen der Statistik. Bern: Huber

Svenson, O.: 1979. Process descriptions of decision making. Organizational Behavior and Human Performance, 23, 86—112

Thorngate, W.: 1980. Efficient Decision Heuristics. Behavioral Science, 25 (3), 219—225

Thorsland, M.N.: 1971. Formative Evaluation in an Audio—tutorial Physics Course with Emphasis on Intuitive and Analytic Problem—solving Approaches, PHD thesis, Cornell University, quoted according to Bastick, 1982

Tietz, R. (ed.): 1983. Aspiration Levels in Bargaining and Economic Decision Making. Berlin, Tokyo: Springer

Tomkins, S.S.: 1962. Affect, imagery, conciousness. Vol. I: The positive affect. New York: Springer

Tomkins, S.S.: 1963. Affect, imagery, conciousness. Vol. II. The negative affect. New York: Springer

Tukey, J.W.: 1977. Exploratory Data Analysis. Reading: Addison — Wesley

Tulving, E.: 1972. Episodic and semantic memory. In: Tulving, E. & Donaldson, W. (eds.), Organization of memory. New York: Academic Press, 381—403

Tulving, E.: 1983. Elements of Episodic Memory. New York: Oxford University Press

Tversky, A.: 1977. Features of Similarity. Psychological Review, 84, 327—352.

Tversky, A. & Kahneman, D.: 1971. Belief in the law of small numbers. Psychological Bulletin, 76(2), 105—110

Tversky, A. & Kahnemann, D: 1973. Availability: A heuristic for judging frequency and probability. Cognitive Psychology, 3, 207–232

Tversky, A. & Kahneman, D.: 1974. Judgment under Uncertainty: Heuristics and biases. Science, 185, 1124–1131

Tversky, A. & Kahneman, D.: 1979. Causal Schemas in Judgments Under Uncertainty. In: Fishbein, M. (ed.), Progress in Social Psychology. Hillsdale: Erlbaum, 49–72

Tversky, A. & Kahneman, D.: 1981. The framing of decisions and the rationality of choice. Science, 211, 453–458

Ungson, G.U. & Braunstein, D.N. (eds.): 1982. Decision Making: An Interdisciplinary Inquiry. Boston: Kent

Varga, T.: 1972. Logic and Probability in the lower grades. In: Educational Studies in Mathematics, 4, 346–357

Wachsmuth, I.: 1981. Zwei Modi der Denktätigkeit – auch beim Mathematik–Lernen? Osnabrücker Schriften zur Mathematik, Osnabrück: FB Mathematik

Wallsten, Th. S.: 1983. The theoretical status of judgmental heuristics. In: Scholz, R.W. (ed.): Decision Making under Uncertainty. Amsterdam: Elsevier (North–Holland), 21–39

Wason, P.C. & Johnson–Laird, P.N.: 1972. Psychology of Reasoning: Structure and Content. Cambridge, Mass.: Harvard University Press

Wason, P.C. & Shapiro, D.: 1971. Natural and contrived experience in a reasoning problem. Quarterly Journal of Experimental Psychology, 23, 63–71

Weiner, B.: 1976. Die Wirkungen von Erfolg und Mißerfolg auf die Leistung. Bern: Huber & Klett

Well, A.D., Pollatsek, A., & Konold, C.E.: 1982. Probability Estimation and the Use and Neglect of Base–Rate Information. Unpublished manuscript, Amherst

Welzel, A.: 1984. Zur Spezialisierung allgemeiner kognitiver Fähigkeiten. Das Affinitätskonzept – entwickelt am Beispiel "Affinität zur Mathematik". In: Reiss, M. & Steiner, H.–G. (eds.), Mathematikkenntnisse, Leistungsmessung, Studierfähigkeit. Köln: Aulis, 170–203

Wertheimer, M.: 1945/1959. Produktives Denken. Frankfurt: Waldemar Kramer. Engl.: Productive Thinking. New York: Harper & Row

Westcott, M.R.: 1968. Toward a Contemporary Psychology of Intuition. New York: Holt, Rinehart & Winston

Westcott, M.R. & Ranzoni, J.H.: 1963. Correlates of intuitive thinking.

Psychological Reports, 12, 595–613
Winer, B.J.: 1971. Statistical principles in experimental design. New York: McGraw Hill
Wishart, D.: 1975. CLUSTAN 1 C user manual. London: Computer center
Wolins, L.: 1982. Research Mistakes. Ames: The Iowa State University Press
Wright, P. & Rip, P.D.: 1981. Retrospective Reports on the Causes of Decisions. Journal of Personality and Social Psychology, 40(4), 601–614
Young, M. & Whitty, G. (eds.): 1977. Society, State, and Schooling. Ringmer: Falmer
Zimmer, A.C.: 1983. Verbal vs. Numerical Processing of Subjective Probabilities. In: Scholz, R.W. (ed.): Decision Making under Uncertainty. Amsterdam: North–Holland, 159–182

INDEX

ACT—star model 145
A priori odds 15
A priori probabilty 14
Abstract vs concrete information
    see availability heuristic
Algebraic stategies in base—rate problems 66, 83, 86—88, 92, 126
ALLAIS' paradox 55
Anchoring and adjustment 90, 91
Analytic
    —algebraic strategies 86—91
    thinking 60, 62, 85—92, 93, 170—184
    mode of thought 62, 84-92, 93, 170—184
    —nonalgebraic strategies 89—72
Aspiration levels 157
Availability heuristic 22, 142, 148, 149

'Base—rate fallacy' 6, 10—21, 94, 98, 133
Base—rate information 13
Base—rate problems
    see Cab problem
    see Hit Parade problem
    see Motor Problem
    see Social Judgment paradigm
    see Tom W. problem
    see TV problem
BAYES' theorem 14—16
BAYES' formula 14—17, 58

Box— and whisker plots 41—42

Cab problem 6, 17, 19—21, 28-—30, 33—34, 59
Calibration tasks 153
Career Socialization and mode of thought 101, 118, 125
Causal schema 22, 142
Central Processor 160—163
Chameleon theory 16
Cluster analyses 76
Cognitive stragegy
    explicit definition 179
Conciousness 162
Concrete vs. abstract information
    see availability heuristic
Conditional probability 58, 85
Confidence and modes of thought 183
Conjunctive probability 58, 84
Cover Framing 105, 114
Credibility of information 159

d_1—measure 37—40
d_2—measure 37—40
Decision Filter 163
Decision making
    under risk 2
    under uncertainty 1
Diagnostic information 14
Diagnosticity information
    see diagnostic information
Diagnosticity response 26, 47, 84, 94, 102, 121, 123, 128, 137

Emotional involvement 64, 85, 177–180
Expert 133
Expert's behavior in base–rate problems 133
Experimental Framing 105, 114
Extremity of base–rates 32–33, 36
Extreme errors
    in base–rate problems 47
Evaluative Structure 157–160
Eye color problem 104–105, 135

Fallacy 21
Frame
    see Framing
Framing 95, 168–169
    see Cover framing
    see Experimental framing
    see Problem framing
    see Problem solving framing
    see Social judgment framing
Framework for the process and structure of human information processing 7, 144

Gestalt 63, 175
Goal System 157–158
Guiding System 160–163

Heuristic Structure 153
Heuristic(s) 21, 23–26, 94, 153, 156
    higher ordered 154, 156
    simple everyday 154, 156
    see Availability heuristic
    see Causal schema
    see Representativeness heuristic
    see Specifity construct
Hit Parade problem 12, 14, 16, 35, 67, 105!, 135

Individualized normative solution 6, 28–30, 50, 142
Individuating information
    see Diagnostic information
Inferential rules 21
Information weights 26–31, 37, 50, 54–56
Interrater reliability 72–74, 111
Introspection 150
Intuition 60–61
    primary 60
    secondary 60
Intuitive
    –algebraic strategies 88–89
    heuristics 100
    mode of thought 60, 62, 84–92, 93, 170–184
    –nonalgebraic stategies 87–88
    predictions 92
    thinking 60, 85–92, 100
    strategies 87–89

Judgmental heuristics 2, 21 , 90, 94, 139
    see Availability heuristic
    see Causal schema
    see Representativeness heuristic
    see Specifity construct

Knowledge
  direct accessible 151–152
  higher ordered 151–152
Knowledge Base 150–153

Light bulb problem 12, 17
Logical probability 19
Long–Term Storage 165

Majority rating 82
Majority rule 155
Microprocesses X, 101
Middling responses 47, 131
Modes of thought 7, 57, 78–92, 93–101, 124, 136, 143, 170–184, 188–189
Motivational aspects of thinking 85
Motor problem 12–14, 16, 35, 67

Natural scientists and mode of thought,
  see Career Socialization
'Normative solution' 21
Objective probability 19

Output 163
Overt Behavior 163

Pedagogical codes 126
Preconcepts of probability 4
Prior odds
  see a priori odds
Prior probability
  see a priori probability
Probability concept
  concept field 185
  logical 19, 58

objective 19, 59
subjective 19, 59
Probability learning 2
Probabilistic reasoning 4
Problem framing 93, 95–98, 101
Problem solving framing 96–98, 101
Prospect theory 95
Protocol analysis 67–69, 94

Radioactive decay 5
Reliability 72–74
Representativeness heuristic 22, 24–26, 142, 155
Roulette problem 104

Satisficing principle 155
Sensory System 146–148
Sex and mode of thought 94, 99, 120, 125, 126
Signal detection theory 148
Social judgment paradigm of 'base-rate fallacy' 11
Social judgment framing 96–98, 101
Sophisticated decision makers 93, 133
Specifity construct 23–26, 53
SPIV–model 167–168
Stand pat choice 32
Statistical inference 4
Stochastic
  different definitions of 3
Stochastic thinking 3, 4!, 64
Subjective probability 19

Textbook paradigm of 'base—rate fallacy 12
Thinking aloud 67—69
Tom W. problem 11, 13, 58, 96
TV problem 12—14, 16—18, 35, 67, 195!

Uncertainty 1—3

Vividness
see Availability heuristic

Wild calculations 91
Working Memory 148—150

z—value for interrater reliability 73—74

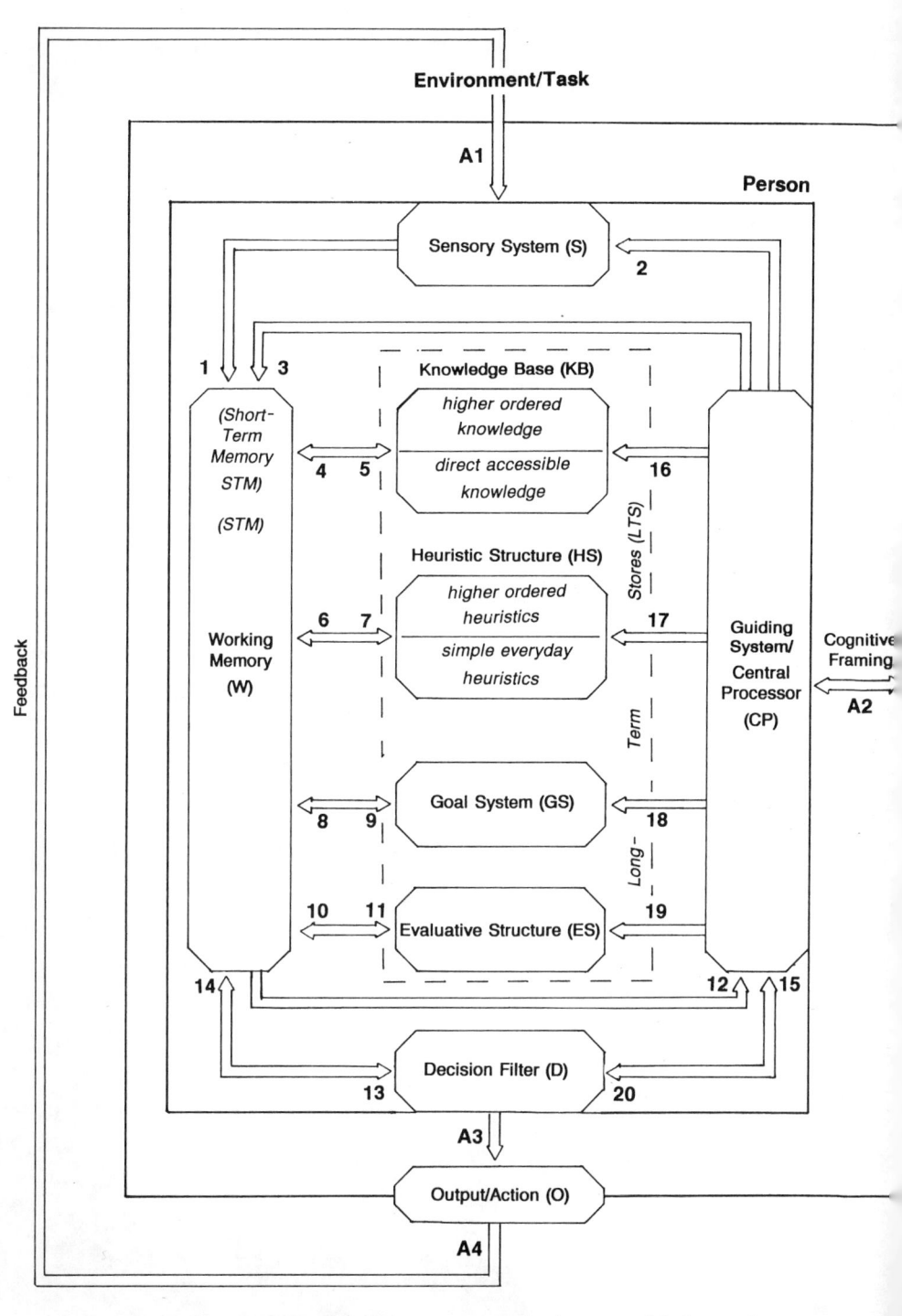

A framework of the structure and process of information processing: The general model